W9-ABY-668

Ivan T. Rathbun

Building Construction Specifications

134

Building Construction Specifications

Building Construction Specifications

IVAN T. RATHBUN
Building Construction, Building Systems Designer

McGRAW-HILL BOOK COMPANY
New York St. Louis San Francisco
Düsseldorf Johannesburg Kuala Lumpur
London Mexico Montreal New Delhi
Panama Rio de Janeiro Singapore
Sydney Toronto

BUILDING CONSTRUCTION SPECIFICATIONS

07-051209-4

123456789EBEB798765432

The editors for this book were Cary Baker and Cynthia Newby, the designer was Allen-Wayne, and its production was supervised by James Lee. It was set in Press Roman by Allen-Wayne.

It was printed and bound by Edwards Brothers, Inc.

CONTENTS

Divisions 1-4 are applicable to all prime contracts.

Divisions 5-34 are applicable to general contracts.

PREFACE

This Text-Workbook uses a totally new approach in teaching specification writing and may be used effectively by students in the classroom, by professionals, or by laymen.

The book contains actual specifications written for buildings already constructed and in operation. They are meant to serve as a guide in preparing specifications under similar circumstances. Every phase of building construction has been covered, including the legal documents and energy systems involved. The preparation of the worksheets will provide actual experience in specification writing.

I wish to express my appreciation here to those architects, engineers, educators, and their societies, who have helped make this book possible.

My thanks also to Alice for her many hours of toil.

INTRODUCTION

How to Use the Building Construction Text-Workbook

Clear, concise specifications are, to quote an old cliché, like having money in the bank. This is true because if the architect, engineer, contractor, owner, and manufacturer all know what is expected of them, costly errors are eliminated. However, to produce this kind of specification, one must first understand what specifications are and what is required.

A dictionary or an encyclopedia would define the word "specifications" as "definite and determinate, as in a contract." As specifications pertain to the design professions, you will find them defined as a "written document, naming and describing equipment, materials, and processes, setting forth therein clear instructions for the execution of that part of the work which cannot be reasonably named, described or shown by graphic illustration." We can conclude then that specifications have two purposes: one, to convey the thought of the designer, his vision of the finished product, to the hundreds of people who will have a hand in the construction of a particular building; and two, to provide a definitive basis on which a legal contract can be executed.

The specification writer, then, must have a full knowledge and understanding of the work to be accomplished and be able to interpret it intelligently so that the people responsible for the execution of the contract can easily understand their obligation.

The first chapters of the Text-Workbook are devoted to the legal documents, or business forms, that are used by many architects and contractors. Some of them are distributed by the American Institute of Architects. Many architects design their own forms, especially the larger firms that have their own computers. These forms, or documents, are completed for you in the text and will serve as guides to help you complete the worksheets.

Divisions 1, 2, 3, and 4, of the Text-Workbook, contain samples of the main conditions of the contract which outline the general business conditions under which the contract will be carried out. To complete the worksheets for these divisions, you will use the text as a guide and fill in the blanks according to the general conditions needed to fulfill the contract on the building or plans that you are originating or using.

In Divisions 5 through 36, the Text-Workbook contains samples of all the conditions, materials, etc., actually needed to construct the building. Also found here are the various energy systems required to permit the building to function properly. It is suggested that you complete the worksheets in these divisions by detailing or specifying the actual conditions, materials, etc., that exist for the building you are designing or are right for the plans you are using.

After completing the entire specification, go back and carefully check it against the plans, and you will see how a good set of specifications can serve as an overall check in preventing errors in design and drafting. Completing the worksheets should remove any doubt as to what a good set of specifications should contain.

NOTICE TO BIDDERS FOR ELEMENTARY SCHOOL

Sealed bids covering the materials and labor required to construct the Proposed (name of building and address) JEFFERSON ELEMENTARY SCHOOL (county) PLATTE (state) IOWA will be received up to (time) 8:00 P.M. (date) 2/12/72 at the (designated location) ARCHITECT'S OFFICE, 821 1ST ST., PLATTE, IOWA. Separate proposals are requested for the following subdivisions of work. General Contractor may submit alternate bid on complete project including the following subdivisions:

a. General
b. Plumbing
c. Electric Heating and Wiring

1. All proposals must be made on the forms provided in the bound copy of the Specifications and Contract Stipulations hereto attached. All proposal must be legibly written in ink or typewritten. No alterations in proposals or in the printed forms will be permitted by erasures or interlineations. Each proposal, in its bound form as furnished by the Architect or Engineer, shall be enclosed in a sealed envelope, addressed to the Owner, and endorsed on the outside with the bidder's name and title of the work, and filed at the Owner's or Architect's or Engineer's office prior to the hour set for opening the bids. Any proposal received after the scheduled closing time for receipt of proposals shall be returned to the bidder unopened.
2. In case of a difference between the stipulated amount of the proposal written in words and the stipulated amount written in figures, the stipulated amount stated in written words shall govern. Should a discrepancy occur between the unit prices and the extended total, the unit price shall govern.
3. Proposals shall be strictly in accordance with the prescribed forms. Proposals carrying riders or qualifications of the bids as submitted may be rejected. The proposals shall be based on the bidder furnishing all of the necessary labor, material, tools and equipment to fully construct the work in accordance with the detailed specifications covering the work.
4. Each proposal must be signed in ink by the bidder with the full name of the bidder and with his full address. In the case of a firm, the name and residence of each member must be inserted, and in case the proposal is submitted by, or in behalf of, a Corporation, it must be signed in the name of such Corporation by an official authorized to bind the bidder.
5. No Bidder may submit more than one proposal. Two or more proposals under different names will not be received from one firm or association.
6. No modification of bids already submitted will be considered unless such modifications are received prior to hour set for opening. Telegraphic modifications will be rejected unless they are confirmed in writing over the signature of the bidder within forty-eight (48) hours of the time set.
7. No Bidder may withdraw a bid for a period of 30 days after the date and hours set for opening. A bidder may withdraw his proposal at any time prior to the expiration of the period during which proposals may be submitted, by written request of the Bidder, which request must be signed in the same manner and by the same person who signed the Proposal.
8. Each bid must be accompanied by a certified check or bidder's bond payable to the Owner in an amount equal to at least 5 percent of the amount of the Proposal. The bid checks of the three low bidders may be retained for a period of not to exceed 15 days, pending the approval of award of contract by the Owner. All other Bidders' checks will be returned immediately after the bids have been tabulated and the three low bids have been determined. Checks which have been retained will be returned when the Bidder, to whom the contract has been awarded, has furnished and filed the necessary number of signed contracts with the Owner, and when the executed contract has been approved by the Owner as to final execution.
9. None of the Information for Bidders, Proposal Form, Contract or Specifications shall be removed from the bound copy of the Contract Documents prior to the filing of same.

10. The Contract will be awarded to the lowest and/or best qualified responsible bidder complying with these instructions and with the Advertisement. Three executed copies of the Contract shall be filed, within ten days after the award, in the office of the Owner.
11. The Contract when executed, shall be deemed to include the entire agreement between the parties thereto, and the Contractor shall not claim any modification thereof resulting from any representation or promise made at any time by any officer, agent, or employee of the Owner or by any other person.
12. The Owner reserves the right to waive minor irregularities or minor errors in any Proposal, if it appears to the Owner that such irregularities or errors were made through inadvertence. Any such irregularities or errors so waived must be corrected on the Proposal in which they occur prior to the execution of any Contract which may be awarded thereon.
13. The Owner reserves the right to reject any or all Proposals.
14. No payment shall be made to the Bidder for materials or labor involved in correcting errors or omissions on the part of the Bidder which result in construction not in accordance with the Specifications.
15. Computation of quantities that will be the basis for payment estimates, both monthly and final, will be made by the Architect.
16. The party to whom the contract is awarded will be required forthwith to execute the contract on the forms which are acceptable to the Owner and the Architect, all within ten calendar days from the date when the written notice of award of the contract is mailed to the Bidder at the address given by him; in case of his failure to do so, the Owner may at his option, consider that the Bidder has abandoned the contract, in which case the certified check or bidder's bond accompanying the proposal shall become the property of the Owner.
17. The Owner will have available all funds necessary for immediate payment of the project.
18. The successful Bidder shall be required to furnish and pay a one hundred percent (100) performance bond for entire contract price. Successful bidder shall be required to furnish and pay for bond for payment of labor and material.

Copies of the Plans and Specifications may be obtained from the Office of (name of architect or engineer) JOHN DOE (address) 821 1ST ST., PLATTE, IOWA.

A deposit of Ten Dollars ($10.00) will be required when Plans and Specifications are removed from place of filing.

Deposit will be returned if Plans and Specifications are returned at time of Bid Letting, or within ten days after Bid Letting.

FORM OF PROPOSAL FOR GENERAL CONTRACT WORK

Gentlemen:

1. The undersigned having visited the site, examined the Plans and Specifications and otherwise familiarized himself with the conditions and requirements necessary and incidental to the completion of the General Contract work for JEFFERSON ELEMENTARY SCHOOL agrees to provide all materials, labor, tools, equipment and services for the following sum:

   Base Bid THREE HUNDRED FIFTY THOUSAND Dollars

   $ 350,000.00

   If regulations, weather and other factors influencing the progress of the work do not materially change from the presently anticipated, we estimate that if awarded the contract we shall complete same in ONE HUNDRED NINETY calendar days from date of contract.

   This base bid proposal may be increased or decreased in accordance with the alternate proposals as may be selected in the following list. Alternates are specified in Division 3 of the Specifications.

   ALTERNATE G-1: This Contractor shall state the amount to add to or deduct from base bid if this Alternate is accepted.

   Add $ 7.500.00 Deduct $ ____

   ALTERNATE G-2: This Contractor shall state the amount to add to or deduct from base bid if this Alternate is accepted.

   Add $ 4,258.00 Deduct $ ____

   ALTERNATE G-3: This Contractor shall state the amount to add to or deduct from base bid if this Alternate is accepted.

   Add $ 1,452.00 Deduct $ ____

   ALTERNATE G-4: This Contractor shall state the amount to add to or deduct from base bid if this Alternate is accepted.

   Add $ 2,484.00 Deduct $ ____

   ALTERNATE G-5: This Contractor shall state the amount to add to or deduct from base bid if this Alternate is accepted.

   Add $ 982.00 Deduct $ ____

2. The undersigned having visited the site, examined the Plans and Specifications and otherwise familiarized himself with the conditions and requirements necessary and incidental to the completion of the General Contract Work, Plumbing and Heating Contract Work and Electrical Contract Work, agrees to provide all materials, labor, tools, equipment and services for the following sum:

   Base Bid FOUR HUNDRED TWENTY FIVE THOUSAND Dollars

   $ 425,000.00

   If regulations, weather and other factors influencing the progress of the work do not materially change from the presently anticipated, we estimate that if awarded the contract we shall complete same in ONE HUNDRED NINETY calendar days from date of contract.

   ALTERNATE E-1: This Contractor shall state the amount to add to or deduct from base bid if this Alternate is accepted.

   Add $ ____ Deduct $ 7,308.00

   a. Accompanying this proposal is a bid security required to be furnished by the Contract Documents, the same being subject to forfeiture in event of default by the undersignee.

b. In submitting this bid it is understood that the right is reserved by the Owner to reject any or all bids, and it is agreed that this bid may not be withdrawn during the period of thirty (30) days provided in the Contract Documents.

All bids shall be sealed and filed with the Architect or Engineer by 8:00 P.M. (time) 2/12/72 (date).

Bid Opening will be 8:00 P.M. (time) 2/12/72 (date) 821 1ST ST., PLATTE, IOWA (location).

FIRM NAME GENERAL CONSTRUCTION CO

BY JOHN DOE

TITLE EXECUTIVE VICE-PRESIDENT

FORM OF PROPOSAL FOR ELECTRICAL CONTRACT WORK

Gentlemen:

1. The undersigned having visited the site, examined the Plans and Specifications and otherwise familiarized himself with the conditions and requirements necessary and incidental to the completion of the Electrical Contract Work for (title) JEFFERSON ELEMENTARY SCHOOL (address) PLATTE (county) PLATTE (state) IOWA agrees to provide all materials, labor, tools, equipment and services for the following sum:

   Base Bid TWELVE THOUSAND Dollars

   $ 12,000.00

   If regulations, weather and other factors influencing the progress of the work do not materially change from the presently anticipated, we estimate that if awarded the contract we shall complete same in ONE HUNDRED NINETY calendar days from date of contract.

   This base bid proposal may be increased or decreased in accordance with the Alternate Proposal if accepted. Alternates are specified in Division 3 of the Specifications.

   ALTERNATE E-1: This Contractor shall state the amount to add to or deduct from base bid if this Alternate is accepted.

   Add $ ______ Deduct $ 7,308.00

2. Accompanying this proposal is a bid security required to be furnished by the Contract Documents, the same being subject to forfeiture in the event of default by the undersignee.
3. In submitting this bid it is understood that the right is reserved by the Owner to reject any or all bids, and it is agreed that this bid may not be withdrawn during the period of thirty (30) days provided in the Contract Documents.

All bids shall be sealed and filed with the Architect or Engineer by 8:00 P.M. (time), 2/12/72 (date).

Bid Opening will be 8:00 P.M. (time) 2/12/72 (date) 821 1ST ST., PLATTE, IOWA (location).

FIRM NAME PLATTE ELECTRIC

BY JOHN PLATTE

TITLE OWNER

FORM OF PROPOSAL FOR PLUMBING CONTRACT WORK

Gentlemen:

1. The undersignee having visited the site, examined the Plans and Specifications and otherwise familiarized himself with the conditions and requirements necessary and incidental to the completion of the Plumbing Contract Work for the (title) __ JEFFERSON ELEMENTARY SCHOOL (address) PLATTE (county) PLATTE (state) IOWA agrees to provide all materials, labor, tools, equipment and services for the following sum:

   Base Bid $ SIXTY THREE THOUSAND Dollars

   $ 63,000.00

   If regulations, weather and factors influencing the progress of the work do not materially change from the presently anticipated, we estimate that if awarded the contract we shall complete same in ONE HUNDRED NINETY calendar days from date of contract.

   This base bid proposal may be increased or decreased in accordance with the following alternate proposal if selected. Alternates are specified in Division 3 of the Specifications.

2. Accompanying this proposal is a bid security required to be furnished by the Contract Documents, the same being subject to forfeiture in event of default by the undersignee.
3. In submitting this bid it is understood that the right is reserved by the Owner to reject any or all bids, and it is agreed that this bid may not be withdrawn during the period of thirty (30) days provided in the Contract Documents.

NOTE: Should the Owner require Performance Bond or Surety Bond of any kind, the premium for such Bond will be paid for by the Owner and the Underwriting Surety will be determined by same.

All bids shall be sealed and filed with the Architect or Engineer by 8:00 P.M. (time) 2/12/72 (date).

Bid Opening will be 8:00 P.M. (time) 2/12/72 (date) 821 1ST ST., PLATTE, IOWA (location).

FIRM NAME MECHANICAL CONTRACTORS

BY JOHN DOE

TITLE OWNER

AGREEMENT

THIS AGREEMENT, made as of the THIRD day of MARCH, 19 72 by and between CITY OF PLATTE hereinafter called the Owner and GENERAL CONSTRUCTION CO. hereinafter called the CONTRACTOR, WITNESSETH, That whereas the OWNER intends to CONTRACT FOR CONSTRUCTION OF JEFFERSON ELEMENTARY SCHOOL hereinafter called the PROJECT, in accordance with plans of JOHN DOE Architect or Engineer, 821 1ST Street, PLATTE, IOWA NOW, THEREFORE, The OWNER and CONTRACTOR for the considerations hereinafter set forth, agree as follows:

Article 1. Scope of Work

The Contractor agrees to furnish all the necessary labor, materials, equipment, tools and services necessary to perform and complete in a workmanlike manner all work required for the construction of the Project, in strict compliance with the Contract Documents herein mentioned which are hereby made part of the Contract, including the following Addenda:

No. 281 Dated 12/12/71

No. 282 Dated 12/16/71

Article 2. Alternates

The following Alternates have been accepted and become a part of the Contract Work: 1, 2, 3, 4, 5

Article 3. Time of Completion

The work to be performed under this contract shall be commenced within 10 calendar days after the Contractor is notified of the approval of the Contract, and shall be completed within ONE HUNDRED NINETY calendar days of the commencement of the Contract Time as defined in the Special Conditions of the Contract.

Article 4. The Contract Sum

The OWNER agrees to pay, and the CONTRACTOR agrees to accept, in full payment for the performance of this Contract, subject to additions and deductions provided therein the Contract amount of:

THREE HUNDRED SIXTY SIX THOUSAND SEVEN HUNDRED SEVENTY SIX Dollars

$ 366,776.00

Article 5. Progress Payments

The OWNER makes progress payments as follows: On or about the tenth day of each month, 85 percent of the value, based on the Contract Prices, labor and materials, incorporated in the work and materials furnished by the Contractor suitably stored at the site thereof up to the first day of the month, as estimated by the Architect or Engineer, less the aggregate of the previous payments, and upon substantial completion of the entire work, a sum sufficient to increase the total payments to 85 percent of the Contract Price.

Article 6. Change Orders – Extra Work

All extra work and changes, alterations, and modifications in the Contract Work and extensions of time will be ordered and/or approved by the OWNER on the standard change order form attached and must be approved by the Architect or Engineer or his representative.

*Article 7. **Waivers***

Waiver of any provisions or specifications or any part of this contract cannot be made without prior recommendation of the Architect or Engineer and written approval of the OWNER. Approval of waivers will be made on the standard change order form.

*Article 8. **Acceptance***

Final inspection and acceptance of the work shall be made by the OWNER, the Architect or Engineer in collaboration with the Contractor. Such inspection shall be made as soon as practical after the contractor has notified the OWNER in writing that the work is ready for inspection.

*Article 9. **Final Estimate and Payment***

Upon the completion and acceptance of the work, the Architect or Engineer shall issue a certificate that the whole work provided for in this contract has been completed and accepted by him under the conditions and terms thereof, and shall make the final estimate of work. Whereupon, the entire balance found to be due the contractor shall be paid to the Contractor by the OWNER in accordance with existing state laws. Before the approval of the final estimate the contractors shall submit evidence satisfactory to the Owner and the Architect or Engineer that all payrolls, material bills, and outstanding indebtedness in connection with this contract have been paid.

The making and acceptance of the final payment shall constitute a waiver of all claims by the Owner, other than those arising from unsettled liens, from faulty work appearing after final payment, or from requirements of the specifications, and of all claims by the Contractor except those previously made and still unsettled.

If, after the work has been substantially completed, full completion thereof is materially delayed through no fault of the Contractor, and the Architect or Engineer so certified, the Owner shall upon certificate of the Architect or Engineer, and without terminating the Contract make payments of the balance due for that portion of the work fully completed and accepted. Such payment shall be made under the terms and conditions governing final payment, except that it shall not constitute a waiver of claims.

*Article 10. **Subcontractors***

The Contractor agrees that the subcontractors listed in the Proposal will not be changed except at the request or with the approval of the Architect or Engineer. The Contractor is responsible to the Owner for the acts and omissions of his subcontractors, and of their direct and indirect employees, to the same extent as he is responsible for the acts and omissions of his employees. The Contract Documents shall not be construed as creating any contractual relation between the Owner and any subcontractor. The Contractor shall bind every subcontractor by the terms of the Contract Documents.

*Article 11. **Contract Documents***

The Contract comprises the Contract Documents listed below. In the event that any provision of one Contract Document conflicts with the provision of another Contract Document, the provision in that Contract Document first listed below shall govern, except as otherwise specifically stated:

a. Agreement (this instrument)
b. Addenda to Contract Documents
c. Remaining Legal and Procedural Documents

1) Proposal
2) Information for Bidders
3) Advertisement

d. Detailed Specification Requirements
e. Drawings
f. Modifications of the General Conditions of the Contract
g. General Conditions of the Contract
h. Bonds
    1) Performance Bonds
    2) Labor and Material Payment Bond
    3) Proposal Guaranty

Article 12. Authority and Responsibilities of the Architect or Engineer

All work shall be done under the general supervision of the Architect or Engineer. The Architect or Engineer shall decide any and all questions which may arise as to the quality and acceptability of Drawings and Specifications and all questions as to the acceptable fulfillment of the Contract on the part of the Contractor.

Article 13. Successors and Assigns

This agreement and all of the covenants thereof shall be inure to the benefit of and be binding upon the Owner and the Contractor respectively and his partners, successors, assigns and legal representatives. Neither the Owner nor the Contractor shall have the right to assign, transfer or sublet his interests or obligations hereunder without written consent of the other party.

Article 15. In Whitness Whereof,

the parties have made and executed this Agreement, the day and year first above written.

PLATTE, IOWA
Owner

By JOHN LARSON

Title MAYOR

Business Address PLATTE, IOWA

GENERAL CONSTRUCTION CO.
Contractor

By JOHN DOE

Title EXECUTIVE VICE-PRESIDENT

Business Address PLATTE, IOWA

ARCHITECT AND ENGINEER'S ASSOCIATION

By JOHN DOE

Title ASSOCIATE

Business Address PLATTE, IOWA

APPLICATION FOR PAYMENT

CONTRACTOR'S APPLICATION NO. 1

ARCHITECT'S JOB NO. JPI-1 PERIOD FROM 3/12/72 TO 3/31/72

TO Platte, Iowa OWNER. APPLICATION IS MADE FOR PAYMENT, AS SHOWN BELOW, IN CONNECTION WITH THE FOOTINGS (CONCRETE) WORK FOR YOUR JEFFERSON ELEMENTARY SCHOOL PROJECT

DESCRIPTION OF WORK	CONTRACT AMOUNT	THIS APPLICATION		COMPLETED		BALANCE TO FINISH
		Labor	Materials	%	TO DATE	
POURED FOOTINGS	12,500.00	4,500.00	8,000.00	90	90	10%

This is to certify that the work as listed above has been completed in accordance with the Contract Documents. That all lawful charges for labor, materials, etc., covered by previous certificates for payment have been paid and that a payment is now due in the amount of

TWELVE THOUSAND FIVE HUNDRED DOLLARS ($ 12,500.00)
from which retainage of 10 % as set out in the Contract Documents shall be deducted 125.00

GENERAL CONSTRUCTION CO. CONTRACTOR

DATE 3/30 19 72 Per JOHN DOE

CERTIFICATE FOR PAYMENT

CERTIFICATE NO. IPI-1 ISSUED: DATE 3/30/72

TO PLATTE, IOWA OWNER

THIS IS TO CERTIFY that in accordance with your contract dated 2/12 19 72,

GENERAL CONSTRUCTION CO. CONTRACTOR

for JEFFERSON ELEMENTARY SCHOOL is entitled to the 1ST payment

which is for the period 3/12, 19 72 through 3/30, 19 72

in the amount of: TWELVE THOUSAND THREE HUNDRED SEVENTY FIVE

Dollars ($ 12,375.00). The present status of the account for the above contract is as follows:

ORIGINAL CONTRACT SUM $ 366,776.00

Change Orders approved in previous months by Owner

	Additions	*Deductions*
Change Order No. NONE	$ NONE	$ NONE
Change Order No. NONE	$ NONE	$ NONE
Change Order No. NONE	$ NONE	$ NONE
Change Order No. NONE	$ NONE	$ NONE
TOTALS	$	$

Total Additions	$ NONE
Sub Total	$
Total Deductions	$ NONE
TOTAL OF CONTRACT TO DATE	$ 12,500.00
Work to finish (this date)	$ NONE
Due Contractor to date	$ 12,500.00
Less retainage ____%	$ 125.00
Total to be drawn to date	$ 12,375.00
Certificates previously issued	$ NONE
THIS CERTIFICATE	$ 12,375.00

This is to certify that all bills are paid for which previous certificates for payment were issued.

CONTRACTOR GENERAL CONSTRUCTION CO.

By JOHN DOE

DATE 3/30, 19 72

CERTIFICATE APPROVAL

OWNER PLATTE, IOWA

BY JOHN LARSON

BY MAYOR

This certificate is based on the estimated amount of work completed in the period covered and any retainage shown is deduced therefrom. This certificate is not negotiable, it is payable only to the payee named herein and its issuance, payment and acceptance are without prejudice to any rights of the OWNER or Contractor under their contract.

ARCHITECT-ENGINEER

By DAVE LOWE

CONTRACT CHANGE ORDER

IOWA
State

JPI-1
Order No.

PLATTE
County

JEFFERSON ELEMENTARY SCHOOL
Contract For

2/2/72
Dated

PLATTE, IOWA
Owner

To GENERAL CONSTRUCTION CO.
(Contractor)

You are hereby requested to comply with the following changes from the contract plans and specifications:

Description of Changes	Decrease in Contract Price	Increase in Contract Price
	$	$
INSTALL FOOTING AND FLAGPOLE		438.00
		438.00
TOTALS	$	438.00
NET CHANGE IN CONTRACT PRICE	$	438.00

Justification: Owner prefers pole separated from building

The sum of $ 438.00 is hereby ADDED TO (added to) (deducted from) the total contract price.

The time provided for completion is NOT CHANGED (increased) (decreased) (not changed) by 190 working days.

This document will become a supplement to the contract and all provisions of the contract will apply hereto.

Requested PLATTE, IOWA, D. LARSON (Owner) 3/30/72 (Date)

Accepted GENERAL CONSTRUCTION CO., JOHN DOE (Contractor) 3/30/72 (Date)

Approved ARCHITECT-ENGINEER, DAVE LOWE (Architect or Engineer) 3/30/72 (Date)

Index to the Articles of the General Conditions

Section 1. Definitions

a. The Contract Documents consist of:
   1) Agreement
   2) Addenda to Contract Documents
   3) Remaining Legal and Procedural Documents
      a) Proposal
      b) Information for Bidders
      c) Advertisement
   4) Detailed Specification Requirements
   5) Drawings
   6) Modifications of the General Conditions of the Contract
   7) General Conditions of the Contract
   8) Bonds
      a) Performance Bond
      b) Labor and Material Payment Bond
      c) Proposal Guaranty

b. The Owner, the Contractor and the Architect or Engineer are those mentioned as such in the Agreement. They are treated throughout the Contract Documents as if each were of the singular number and masculine gender.

c. Wherever in this Contract the word Architect or Engineer is used it shall be understood as referring to the Architect or Engineer of the Owner, acting personally or through an assistant duly authorized in writing for such act by the Architect or Engineer.

d. Written notice shall be deemed to have been duly served if delivered in person to the individual, or to a member of the firm or to an officer of the Corporation for whom it is intended, or if delivered at or sent by registered mail to the last business address known to him who gives the notice.

e. The term Subcontractor, as employed herein, includes only those having a direct contract with the Contractor and it includes one who furnished material worked to a special design according to the plans or specifications of this work, but does not include one who merely furnishes material not so worked.

f. The term "work" of the Contractor or Subcontractor includes labor or materials or both, equipment, transportation, or other facilities necessary to complete the Contract.

g. All time limits stated in the Contract Documents are of the essence of the Contract.

h. Working days are defined as those days when the temperature is above the minimum required for the work being done, but not including holidays, Sundays, or those days when the work could not be done because of inclement weather.

i. Whenever it is permitted in these specifications to submit the "equal" of a designated brand or grade of materials or workmenship, it is understood that in such case the written approval of the Architect for such substitution must be obtained especially for this work.

Section 2. Execution, Correlation and Intent of Documents

The Contract Documents shall be signed in triplicate by the Owner and the Contractor. In case the Owner and the Contractor fail to sign the General Conditions, Drawings or Specifications, the Architect shall identify them.

The Contract Documents are complimentary, and what is called for by any one shall be as binding as if called for by all. The intention of the documents is to include all labor and materials, equipment and transportation necessary for the proper execution of the Work. It is not intended, however, that materials, or work not covered by or properly inferable from any heading, branch, class or trade of the specifications shall be supplied unless distinctly so noted on the drawings. Materials or work described in words which so applied have a well known technical or trade meaning shall be held to refer to such recognized standards.

Section 3. Detail Drawings and Instructions

The Architect or Engineer shall furnish, with reasonable promptness, additional instructions, by means of drawings or otherwise, necessary for the proper execution of the work. All such drawings and instructions shall be consistent with the Contract Documents, true developments thereof, and reasonably inferable therefrom.

Section 4. Copies of Drawings Furnished

Unless otherwise provided in the Contract Documents, the Architect or Engineer shall furnish to the Contractor, free of charge, all copies of drawings and specifications reasonably necessary for the execution of the work.

Section 5. Order of Completion

The Contractor shall submit, at such times as may be requested by the Architect or Engineer, schedules which shall show the order in which the Contractor proposes to carry on the work with dates at which the Contractor will start the several parts of the work and estimated dates of completion of the several parts.

Section 6. Drawings and Specifications on the Work

The Contractor shall keep one copy of all drawings and specifications on the work in good order and available to the Architect or Engineer and to his representatives.

Section 7. Ownership of Drawings

All drawings, specifications and copies thereof furnished by the Architect or Engineer are his property. They are not to be used on other work, and, with the exception of the signed contract set, are to be returned to him on request, at the completion of the work. All models are the property of the Owner.

Section 8. Contractor's Understanding

It is understood and agreed that the Contractor has, by careful examination, satisfied himself as to the nature and location of the work, the conformation of the ground, the character, quality and quantity of the materials to be encountered, the character of equipment and facilities needed preliminary to and during the prosecution of the work, the general and local conditions, and all other matters which can in any way affect the work under this Contract. No verbal agreement or conversation with any officer, agent or employee of the Owner either before or after the execution of this Contract, shall affect or modify any of the terms or obligations herein contained.

Section 9. Materials, Appliances, Employees

Unless otherwise stipulated, the Contractor shall provide and pay for all materials, labor, water, tools, equipment, light, power, transportation and other facilities necessary for the execution and completion of the work.

Unless otherwise specified, all materials shall be new and both workmanship and materials shall be of a good quality. The Contractor shall, if required, furnish satisfactory evidence as to the kind and quality of materials.

The Contractor shall at all times enforce strict discipline and good order among his employees, and shall not employ on the work any unfit person or anyone not skilled in the work assigned to him.

Section 10. Royalties and Patents

The Contractor shall pay all royalties and license fees. He shall defend all suits or claims for infringement of any patent rights and shall save the Owner harmless from loss on account thereof, except that the Owner shall be responsible for all such loss when a particular process or the product of a particular manufacturer or manufacturers is specified, but if the Contractor has information that the process or article specified is an infringement of a patent he shall be responsible for such loss unless he promptly gives such information to the Architect or Engineer.

Section 11. Surveys, Permits and Regulations

The owner shall furnish all surveys unless otherwise specified. Permits and licenses of a temporary nature necessary for the prosecution of the work shall be secured and paid for by the Contractor. Permits, licenses, and easements for permanent structures or permanent changes in existing facilities shall be secured and paid for by the Owner, unless otherwise specified.

The Contractor shall give all notices and comply with all laws, ordinances, rules and regulations bearing on the conduct of the work as drawn and specified. If the Contractor observes that the drawings and specifications are at variance therewith, he shall promptly notify the Architect or Engineer in writing, and any necessary changes shall be adjusted as provided in the Contract for changes in the work. If the Contractor performs any work knowing it to be contrary to such laws, ordinances, rules and regulations, and without such notice to the Architect, he shall bear all costs arising therefrom.

Section 12. Protection of Work and Property

The Contractor shall continuously maintain adequate protection of all his work from damage and shall protect the Owner's property from injury or loss arising in connection with this Contract. He shall make good any such damage, injury or loss, except such as may be directly due to errors in the Contract Documents or caused by agents or employees of the Owner. He shall adequately protect adjacent property as provided by law and the Contract Documents. He shall provide and maintain all passageways, guard fences, lights and other facilities for protection required by public authority or local conditions.

In an emergency affecting the safety of life or of the work or of adjoining property, the Contractor, without special instructions or authorization from the Architect or Engineer is hereby permitted to act, at his discretion, to prevent such threatened loss or injury, and he shall so act, without appeal, if so instructed or authorized. Any compensation, calimed by the Contractor on account of emergency work, shall be determined by agreement or arbitration.

Section 13. Inspection of Work

The Architect or Engineer and his representatives shall at all times have access to the work wherever it is in preparation or progress and the Contractor shall provide facilities for such access and for inspection.

If the specifications, the Architect's or Engineer's instructions, laws, ordinances or any public authority require any work to be specially tested or approved, the Contractor shall give the Architect or Engineer timely notice of its readiness for inspection, and if the inspection is by another authority than the Architect or Engineer, of the date fixed for such inspection. Inspections by the Architect or Engineer shall be promptly made, and where practicable at the source of supply. If any work should be covered up without approval of consent of the Architect, or Engineer, it must, if required by the Architect or Engineer, be uncovered for examination at the Contractor's expense.

Re-examinations of questioned work may be ordered by the Architect or Engineer and if so ordered the work must be uncovered by the Contractor. If such work be found in accordance with the Contract Documents the Owner shall pay the cost of re-examination and replacement. If such work be found not in accordance with the Contract Documents the Contractor shall pay such cost, unless he shall show that the defect in the work was caused by another Contractor, and in that event the Owner shall pay such cost.

Section 14. Superintendence: Supervision

The Contractor shall keep on his work during its progress a competent superintendent and any necessary assistants, all satisfactory to the Architect or Engineer. The superintendent shall not be changed except with the consent of the Architect or Engineer, unless the superintendent proves to be unsatisfactory to the Contractor and ceases to be in his employ. The superintendent shall represent the Contractor in his absence and all directions given to him shall be as binding as if given to the Contractor. Important directions shall be confirmed in writing to the Contractor. Other directions shall be so confirmed on written request in each case. The Contractor shall give efficient supervision to the work, using his best skill and attention.

If the Contractor, in the course of the work, finds any discrepancy between drawings and the physical conditions of the locality, or any errors or omissions in drawings or in the layout as given by points and instructions, it shall be his duty to immediately inform the Architect or Engineer, in writing, and the Architect shall promptly verify the same. Any work done after such discovery until authorized, will be done at the Contractor's risk.

Neither party shall employ or hire any employee of the other party without his consent.

Section 15. Changes in the Work

The Owner, without invalidating the Contract, may order extra work or make changes by altering, adding or deducting from the work, the Contract Sum being adjusted accordingly. All such work shall be executed under the conditions of the original Contract except that any claim for extension of time caused thereby shall be adjusted at the time of ordering such change.

In giving instructions, the Architect or Engineer shall have authority to make minor changes in the work, not involving extra cost, and not inconsistent with the purposes of the work, but otherwise, except in an emergency endangering life or property, no extra work or change shall be made unless in pursuance of a written order by the Architect or Engineer, and no claim for an addition to the Contract Sum shall be valid unless so ordered.

The value of any such work or change shall be determined in one or more of the following ways:

a. By estimate and acceptance in a lump sum
b. By unit prices named in the Contract or subsequently agreed upon
c. By cost and percentage or by cost and a fixed fee

If none of the above methods is agreed upon, the Contractor, provided he receives an order as above, shall proceed with the work. In such case and also under case c., he shall keep and present in such form as the Architect or Engineer may direct, a correct account of the net cost of labor and materials, together with vouchers. In any case, the Architect or Engineer will certify to the amount, including reasonable allowance for overhead and profit, due to the Contractor. Pending final determination of value, payments on account of changes shall be made on the Architect's or Engineer's estimate.

Section 16. Extension of Time

Extension of time stipulated in the Contract for completion of the work will be made if and as the Architect or Engineer may deem proper; when work under extra work order as hereinbefore provided is added to the work under this Contract: when the work is suspended as provided in Section 20: and when the work of the Contractor is delayed on account of conditions which could not have been foreseen, or which were beyond the control of the Contractor, and which were not the result of his fault or negligence. Extension of time for completion will also be allowed for any delays in the progress of the work caused by any act or neglect of the Owner or of his employees or by other Contractors employed by the Owner, or delay due to an act of government, or by any delay in the furnishing of plans and necessary information by the Architect or Engineer, or by any other cause which in the opinion of the Architect or Engineer entitles a Contractor to an extension of time. Strikes and other labor disputes shall also be cause for an extension of time.

The Contractor shall notify the Architect or Engineer promptly of any occurrence of conditions which in the Contractor's opinion entitle him to an extension of time. Such notice shall be in writing and shall be submitted in ample time to permit full investigation and the validation of the Contractor's claim. The Architect or Engineer shall acknowledge receipt of the Contractor's notice within five days of its receipt. Failure to provide such notice shall constitute a waiver by the Contractor of any such claim.

Section 17. Claims for Extra Cost

If the Contractor claims that any instructions by drawings or otherwise involve extra cost under this Contract, he shall give the Architect or Engineer written notice thereof within ten days after the receipt of such instructions and in any event before proceeding to execute work, except in emergency endangering life or property, and the procedure shall then be as provided for changes in the work. No such claim shall be valid unless so made.

Section 18. Deductions for Uncorrected Work

If the Architect or Engineer deems it inexpedient to correct work injured or done not in accordance with the Contract, an equitable deduction from the Contract price shall be made therefor.

Section 19. Delays and Extension of Time

If the Contractor be delayed at any time in the progress of the work by any act or neglect of the Owner, or of his employees, or by any other Contractor employed by the Owner, or by changes ordered in the work, or by strikes, lockouts, fire, unusual delay in transportation, unavoidable casualties or any causes beyond the Contractor's control, or by delay authorized by the Architect or Engineer pending arbitration, or by any cause which the Architect or Engineer shall decide to justify the delay, then the time of completion shall be extended for such reasonable time as the Architect or Engineer may decide.

No such extension shall be made for delay occurring more than seven days before claim therefor is made in writing to the Architect or Engineer. In the case of a continuing cause of delay only one claim is necessary.

If no schedule or agreement stating the dates upon which drawings shall be furnished is made, then no claim for delay shall be allowed on account of failure to furnish drawings until two weeks after demand for such drawings and not then unless such claim be reasonable.

This article does not exclude the recovery of damages or delay by either party under other provisions in the Contract Documents.

Section 20. Correction of Work Before Final Payment

The Contractor shall promptly remove from the premises all materials condemned by the Architect or Engineer as failing to conform to the Contract, whether incorporated in the work or not, and the Contractor shall promptly replace and re-execute his own work in accordance with the Contract and without expense to the Owner and shall bear the expense of making good all work of other contractors destroyed or damaged by such removal or replacement.

If the Contractor does not remove such condemned work and materials within a reasonable time, fixed by written notice, the Owner may remove them and may store the material at the expense of the Contractor. If the Contractor does not pay the expense of such removal within ten days' time thereafter, the Owner may, upon ten days' written notice, sell such materials at auction or at private sale and shall account for the net proceeds thereof, after deducting all the costs and expenses that should have been borne by the Contractor.

Section 21. Suspension of Work

The Owner may at any time suspend the work, or any part thereof by giving ten days' notice to the Contractor in writing. The work shall be resumed by the Contractor within ten (10) days after the date fixed in the written notice from the Owner to the Contractor to do so. The Owner shall reimburse the Contractor for expense incurred by the Contractor in connection with the work under this Contract as a result of such suspension.

But if the work or any part thereof shall be stopped by the notice in writing aforesaid, and if the Owner does not give notice in writing to the Contractor to resume work at a date within thirty (30) days of the date fixed in the written notice to suspend, then the Contractor may abandon that portion of the work so suspended and he will be entitled to the estimates and payments for all work done on the portions so abandoned, if any.

Section 22. The Owner's Right to do Work

If the Contractor shall neglect to prosecute the work properly or fail to perform any provision of this Contract, the Owner, after three days' written notice to the Contractor, may, without prejudice to any other remedy he may have, made good such deficiencies and may deduct and delete the cost thereof from the payment then or thereafter due the Contractor.

Section 23. The Owner's Right to Terminate Contract

If the Contractor should be adjudged a bankrupt, or if he should make a general assignment for the benefit of his creditors, or if a receiver should be appointed on account of his insolvency, or if he should persistently or repeatedly refuse or should fail, except in cases for which extension of time is provided, to supply enough properly skilled workmen or proper materials, or if he should fail to make prompt payment to subcontractors or for material or labor or persistently disregard laws, ordinances or the instructions of the Architect or Engineer, or otherwise be guilty of a substantial violation of any provision of the contract, then the Owner, upon the certificate of the Architect or Engineer that sufficient cause exists to justify such action, may, without prejudice to any other right or remedy and after giving the Contractor seven days' written notice, terminate the employment of the Contractor and take possession of the premises and of all materials, tools and appliances thereon and finish the work by whatever method he may deem expedient. In such a case the Contractor shall not be entitled to receive any further payment until the work is finished. If the unpaid balance of the contract price shall exceed the expense of finishing the work, including compensation for additional managerial and administrative services, such excess shall be paid to the Contractor. If such expense shall exceed such unpaid balance, the Contractor shall pay the difference to the Owner. The expense incurred by the Owner as herein provided, and the damage incurred through the Contractor's default, shall be certified by the Architect or Engineer.

Section 24. Contractor's Right to Stop Work or Terminate Contract

If the work should be stopped under an order of any court, or other public authority, for a period of three months, through no act or fault of the Contractor or of anyone employed by him, or if the Architect or Engineer should fail to issue any estimate for payment within seven days after it is due, or if the Owner should fail to pay the Contractor within seven days of its maturity and presentation, any sum certified by the Architect or Engineer or awarded by aribitrators, then the Contractor may, upon seven days' written notice to the Owner and the Architect or Engineer, stop work or terminate this Contract and recover from the Owner payment for all work executed and any loss sustained upon any plant or materials and reasonable profit and damages.

Section 25. Removal of Equipment

In the case of annulment of this Contract before completion from any cause whatever, the Contractor, if notified to do so by the Owner, shall promptly remove any part or all of his equipment and supplies from the property of the Owner, failing which the Owner shall have the right to remove such equipment and supplies at the expense of the Contractor.

Section 26. Use of Completed Portions

The Owner shall have the right to take possession of and use any completed or partially completed portions of the work, notwithstanding the time for completing the entire work or such portions that may not have expired, but such taking possession and use shall not be deemed an acceptance of any work not completed in accordance with the Contract Documents. If such prior use increases the cost of or delays the work, the Contractor shall be entitled to such extra compensation, or extension of time, or both, as the Architect or Engineer may determine.

Section 27. Responsibility for Work

The Contractor assumes full responsibility for materials and equipment used in the construction of the work and agrees to make no claims against the Owner for damages to such materials and equipment from any cause except negligence or willful act of the Owner. Until its final acceptance, the Contractor shall be responsible for damage to or destruction of the project (except such work covered by partial acceptance as set forth in Section 26). He shall make good all work damaged or destroyed before acceptance.

Section 28. Payments Withheld–Prior to Final Acceptance of Work

The Owner may withhold or, on account of subsequently discovered evidence, nullify the whole or a part of any certificate to such extent as may be necessary to protect himself from loss on account of:

a. Defective work not remedied
b. Claims filed or reasonable evidence indicating probable filing of claims
c. Failure of the Contractor to make payments to subcontractors or for material or labor
d. A reasonable doubt that the Contract can be completed for the balance then unpaid
e. Damage to another Contractor

When the above grounds are removed or the Contractor provides the Surety Bond satisfactory to the Owner, which will protect the Owner in the amount withheld, payment shall be made from amounts withheld, because of them.

Section 29. Contractor's Liability Insurance

The Contractor shall secure and maintain such insurance policies as will protect himself, his subcontractors, and unless specified, the Owner, from claims from bodily injuries, death or property damage which may arise from operations under this Contract, whether such operations be by himself or by any subcontractor or anyone employed by them directly or indirectly. The following insurance policies are required:

a. Statutory Workman's Compensation
b. Bodily Injury–Each person $50,000.00; Each Accident $100,000.00; Property Damage; Each accident $50,000.00
c. Automobile Public Liability and Property Damage–Bodily Injury; Each person $50,000.00 Each accident $100,000.00; Property Damage; Each accident $50,000.00

Certificates of such insurance shall be filed with the Architect or Engineer or Owner, and shall be subject to approval as adequacy and protection. Said certificates of insurance shall contain a ten day notice of cancellation.

Section 30. Surety Bonds

Contractor shall be required to furnish and pay for one hundred percent (100%) Performance Bond for entire contract price and also furnish and pay for bond for payment of labor and materials.

Section 31. Damage Claims

The Contractor shall defend, indemnify and save harmless the Owner, its officers, agents, servants, and employees against and from all suits, losses, demands, payments, actions, recoveries, judgments and costs of every kind and description, and from all damages to which the Owner or any of its officers, agents, servants and employees may be subjected by reason of injury to the personal property of others resulting from the performance of the project or through any improper or defective machinery, implements or appliances used by the Contractor in the project, or through any act or omission on the part of the Contractor on his agents, employees, or servants; that he shall further defend, indemnify and save harmless the Owner, its officers, agents, servants and employees from all suits and actions of any kind or character whatsoever which may be brought or instituted by any subcontractor, materialsman or laborer, who has performed work or furnished materials in or about the project or by, or on account of, any claims or amount recovered for an infringement or patent, trademark or copyright.

Section 32. Liens

Neither the final payment nor any part of the retained percentage shall become due until the Contractor, if required, shall deliver to the Owner a complete release of all liens arising out of this Contract, or receipts in full in lieu thereof and, if required in either case, an affidavit that so far as he has knowledge or information the releases and receipts include all the labor and material for which a lien could be filed; but the Contractor may, if any subcontractor refuses to furnish a release or receipt in full, furnish a bond satisfactory to the Architect or Engineer, to indemnify the Owner against against any lien. If any lien remains unsatisfied after all payments are made, the Contractor shall refund to the Owner all moneys that the latter may be compelled to pay in discharging such a lien, including all costs and a reasonable attorney's-fee.

Section 33. Assignment

Neither party to the Contract shall assign the Contract or sublet it as a whole without the written consent of the other, nor shall the Contractor assign any money due or to become due to him hereunder without the previous written consent of the Architect or Engineer.

Section 34. Rights of Various Interests

Wherever work being done by the Owner's forces or by other contractors is contiguous to work covered by this Contract the respective rights of the various interests involved shall be established by the Architect or Engineer, to secure the completion of the various portions of the work in general harmony.

Section 35. Separate Contracts

The Owner reserves the right to let other contracts in connection with this work. The Contractor shall afford other contractors reasonable opportunity for the introduction and storage of their materials and the execution of their work, and shall properly connect and coordinate his work with theirs.

If any part of the Contractor's work depends for proper execution of results upon the work of any other Contractor, the Contractor shall inspect and promptly report to the Architect or Engineer any defects in such work that render it unsuitable for such proper execution and results. His failure so to inspect and report shall constitute an acceptance of the other contractor's work as fit and proper for the reception of his work, except as to defects which may develop in the other contractor's work after the execution of his work.

To insure the proper execution of his subsequent work the Contractor shall measure work already in place and shall at once report to the Architect or Engineer any discrepancy between the executed work and the drawings.

Section 36. Subcontracts

The Contractor shall, as soon as practicable after the signature of the Contract, notify the Architect or Engineer in writing of the names of subcontractors proposed for the work and shall not employ any that the Architect or Engineer may within a reasonable time object to as incompetent or unfit.

The Contractor agrees that he is as fully responsible to the Owner for the acts and omissions of his subcontractors and of persons either directly or indirectly employed by them, as he is for the acts and omissions of persons directly employed by him.

Nothing contained in the Contract Documents shall create any contractual relation between any subcontractor and the Owner.

Section 37. Architect's or Engineer's Status

The Architect or Engineer shall have general supervision and direction of the work. He has authority to stop the work whenever such stoppage may be necessary to insure proper execution of the Contract. He shall also have authority to reject all work and materials which do not conform to the Contract, to direct the application of forces to any portion of the work, as in his judgment is required, and to order the force increased or diminished, and to decide questions which arise in the execution of the work.

Section 38. Architect's or Engineer's Decisions

The Architect or Engineer shall, within a reasonable time after their presentation to him, make decisions in writing on all claims of the Owner or the Contractor and on all other matters relating to the execution and progress of the work or the interpretation of the Contract Documents.

All such decisions of the Architect or Engineer shall be final except in cases where time and/or financial consideration are involved, which, if no agreement in regard thereto is reached, shall be subject to arbitration.

Section 39. Arbitration

Demand for Arbitration. Any and all disputes arising out of, under, or in connection with the contract or for a breach thereof, shall be submitted to arbitration in accordance with the rules of the American Arbitration Association, Inc., upon demand of either party to the dispute.

Section 40. Lands for Work

The Owner shall provide as indicated on drawings and not later than the date when needed by the Contractor the lands upon which the work under this Contract is to be done, rights-of-way for access to same, and such other lands which are designated on the darwing for the use of the Contractor. Any delay in the furnishing of these lands by the Owner shall be deemed proper cause for an equitable adjustment in both Contract price and time of completion.

The Contractor shall provide at his own expense, and without liability to the Owner any additional land and access thereto, as may be required for temporary construction facilities, or for storage of materials.

Section 41. Cleaning Up

The Contractor shall, as directed by the Architect or Engineer, remove at his own expense from the Owner's property and from all public and private property all temporary structures, rubbish and waste materials resulting from his operations.

Section 42. Payments

Partial payments will be made on monthly estimates rendered by the Contractor and approved by the Architect or Engineer. The Contractor shall submit his estimate to the Architect or Engineer at least five (5) days prior to the first of the month. This estimate shall be broken down in accordance with the itemized complete cost shown in Proposal forms.

After final completion of the work, the Architect or Engineer shall make a certificate of completion recommending that the work be approved and the Contractor paid.

1. Time of Completion

The Contractor shall commence work within five (5) days of written notice to proceed. The number of calendar days required to complete the work shall be stated in the bidder's proposal.

2. Examination of Site

Before bidding the work each contractor shall inform himself fully as to all site conditions.

3. Laying Out Work

The Contractor shall at his own expense employ an engineer to give the required lines, points and levels.

4. Temporary Heat

Should the completion of the building proceed into cold weather the Owner will permit the use of the heating plant, or such portion of it as may have been installed, by the General Contractor at his own risk and responsibility. All direct material, fuel, and labor costs at the project in connection with such damage to the heating plant or the building shall be assumed by him

5. Temporary Utilities

a. *Water.* Contractor shall pay for water used for his operations.
b. *Electrical.* General Contractor must pay for electricity used on the job. Electrical Contractor, if required, shall provide temporary connections for the heating–cooling equipment.

6. Temporary Toilets

The General Contractor shall provide and maintain a toilet for the use of workmen. When the job is complete, toilet and waste shall be disposed of in a manner agreeable to the Architect or Engineer.

7. Scope of Work and Work not Included in Construction Contract

The work to be included in this contract includes all labor and material necessary for and reasonably incidental to the erection of the new building, complete as shown and specified in the drawings and specifications.

8. Storage of Materials

The Contractor may store materials on the site. The Contractor is responsible for all material until building completion and possession by Owner.

9. Safety Requirements

Precautions shall be exercised at all times for the protection of persons and property.

10. Samples Required

Samples are required as specified under the various divisions of the work.

The Contractor shall state on the proposal form the amount to add to or subtract from the base bid if the following alternates are to be accepted.

ALTERNATE G-1: Substitute steel roof deck over the lower portion of the roof (not including the multi-purpose room roof) in lieu of Petrel or specified; see Drawing G-1 and Divisions 15 and 23 of the specifications.

ALTERNATE G-2: Add Folding tables, wall recesses and metal frames as detailed on drawings and specified in Division 33. Location of tables is shown on the floor plan.

ALTERNATE G-3: Add the canopy at the main entrance as shown on the drawings and specified in Division 33.

ALTERNATE G-4: Add borrow lights in the corridors as detailed on the drawing and specified in Division 24.

ALTERNATE G-5: Add backstops as specified in Division 33.

ALTERNATE E-1: This contractor shall submit with his bid the amount of deduction to omit the program system and all other clocks from the contract. A conduit shall be installed.

1. General Notes

a. It shall be the responsibility of this Contractor to remove the existing building on the present site.

b. Such building shall be demolished and all materials removed from the job site.

c. No material from the demolished building shall be used in the new construction, either for rough work, form work or any other type of construction, to prevent possible infestation of termites in the new building.

d. All materials in the demolished building shall belong to the Contractor, and it shall be his right to dispose of such materials as he sees fit.

e. Any money received by the Contractor as the result of sale of any parts, materials, and components shall be retained by the Contractor.

f. The Contractor shall include the cost of demolition of the building in his base bid. This amount shall include the disconnect of all utilities, and the removal of any such utilities that will hinder the construction of the new project. If any utilities from the existing building are not removed, they shall be disconnected at the service supply and the unused portions of the conduit pipes shall be blanked off at the service source and load end.

1. General Notes

a. The Contractor shall perform all clearing, stripping, cutting, and rough and finish grading as indicated on the drawings and as hereinafter described. The Contractor shall, as a part of his contract, perform all trenching for footings, all backfill against walls, and all fill under concrete floors, porches, or platforms resting directly on the ground.
b. Excavating, backfilling for sewers, water, gas, plumbing, heating, and electrical work are to be included under their respective divisions.
c. Remove and dispose off site unsuitable and excess material.
d. Seeding and sodding not included in this contract.

2. Subsurface Soil Data (See also Special Conditions)

If any serious obstacles or conditions are encountered, the Architect or Engineer shall be notified immediately and adjustments or alternates worked out.

3. Excavating and Filling

a. Excavate to the dimensions shown on the various drawings, and do all other excavation necessary to fully carry out the work shown or herein specified. Provide shoring wherever necessary. All footing trenches are to be excavated to the exact depths and widths required, with the sides plumb and the bottom level. In no case shall bottom of footing be less then 1'–6" below natural grade line and 3'–0" below finish grade line.
b. The contractor shall make such excavation and provide such drainage ditches as will at all times keep the building site free from standing or running water and will be held fully responsible for the consequences or failure to do so.
c. Filled areas under concrete slabs are to be "filled" with gravel and sand.
d. Where trenches for footings are not straight and true, wood forming will be required. Attention must be given to the top edge of the trench to prevent cave-ins.
e. Gravel-sand fill shall be placed in 6" layers and shall be uniformly moist to permit compaction with a 200 lb cylindrical roller and mechanical tamper. Gravel-sand sample for use in "filled" areas shall be approved by the Architect or Engineer.
f. If suitable bearing for foundations is not encountered at depth indicated, contractor to immediately notify Architect or Engineer and not proceed further until instructions given.
g. Care must be taken to prevent damage to existing sewage drainage tile and water supply pipe on existing site. Work on the existing system is to be accomplished by the mechanical constractor. See site plan for approximate location of piping and tile systems. Necessary measurements made to establish addition of volume of excavation.
h. Protect bottom of excavation from frost. No foundations or slabs on frozen ground.
i. Fill excess cuts under footings with concrete.

4. Grading

a. Grading of the site in general, shall be done by the Contractor. Soil taken from the building, excavation and from the stripping operations shall be used in grading as herein specified, to the grades and limits indicated on the plot plan. Provide additional earth for grading if necessary.
b. Rough grading shall be completed as soon as practicable but final or finish grading shall be delayed until construction is nearing completion.
c. Remove debris from excavations before backfilling; backfill must be free from plaster, bolts, and other debris. Deposit in layers not exceeding 8" under slabs, pavements or other surface; 12" other areas; compact each layer fill uniformly on both sides of foundation walls.

1. General Notes

a. The Contractor shall run an earth test to a depth as necessary to obtain a solid point of earth for establishing firm foundation of structure.
b. The Contractor shall include the cost of this material and operation in his bid.
c. Piles shall be placed a minimum of four feet apart.
d. Piles shall be constructed of steel and as recommended by the structural engineers or the civil engineers completing the soil test. Such expense for all of this work shall be paid by the Owner with an additional fee of eight percent of the total cost of the work completed for testing.
e. All information of testing, and recommendations from structural or civil engineers shall become the property of the Owner.
f. The related costs of these operations are not a part of the building costs.

1. Work Included

a. All structural concrete, footings, slabs, spash blocks, lintels, lintel blocks, filler materials, sidewalks and curbs.

2. Material

a. *Portland Cement,* ASTM C150, Type 1.
b. *Coarse Aggregate,* Hard, durable, uncoated crushed stone or gravel 95 to 100% to pass a 1½" sieve, 35 to 70% to pass a ¾" sieve; 10 to 30% to pass a 3/8" sieve and not over 5% to pass a No. 4 sieve, percentage by weight.
c. *Sand*–Hard, durable, uncoated grains free from salt, loam and clay, fine to coarse with 95 to 100% by weight passing No. 4 sieve; 46 to 70% passing No. 16 sieve; 15 to 30% passing No. 50 sieve and 3 to 8% passing No. 100 sieve.
d. *Sand–Gravel*–Pit run sand gravel washed aggregate mixture meeting approval of Architect or Engineer may be used in lieu of separate sand and gravel. Composed of clean, hard, durable pebbles, free from flaky particles.
e. *Mixing Water*–Suitable for drinking.
f. *Ready Mix Concrete*–May be used if it contains not less than 6 sacks of cement per cubic yard and is a proportioned mix guaranteed by the supplier (ready mix company) to develop a minimum average compressive strength of 1800 pounds per square inch at the end of 7 days and 3000 pounds at the end of 28 days.
g. *Expansion Joint Filler*–Premolded fiberboard impregnated 35 to 50 percent asphalt by weight, full thickness of slot or joint, ½" thick or as directed.

3. Storage of Materials

All cement for this work shall be delivered at least three days before using to allow for testing, if so desired, and shall be kept stored in a safe and dry place. Sand or crushed rock will not be permitted to be piled directly upon the earth, and must be placed upon a platform of wood planks. Any cement found to be lumpy or otherwise injured by age or moisture will be rejected.

4. Construction Forms

a. Plumb and straight. Brace to prevent displacement, no form coatings to strain exposed concrete.
b. Footing forms may be omitted if excavation to neat sizes and soil condition approved.

5. Strength and Proportions of Concrete

a. All structural concrete shall contain not less than 6 sacks of cement per cubic yard and shall develop a minimum average compressive strength of 1800 pounds per square inch at the end of 7 days and 3000 pounds at the end of 28 days.
b. Cylinders for compression tests shall be taken at each of four different locations in footing pour. Furnish three cylinders for each test; one for 7 day test, one for 28 day and one for spare.
c. Compression tests shall be made by a recognized testing laboratory and costs of tests shall be paid for by Contractor.
d. Slump test according to ASTM Specification C-143 not more than 3".

6. Consistency of Concrete Work

Enough water shall be used to produce concrete that will flow into forms and about the reinforcement and which will not separate the aggregate from the mortar during wheeling or placing. Water shall be kept to the minimum possible amount so that when placed, no excess water will be brought to the surface. Use no more than 6½ gallons of water per sack of cement.

7. *Protection of Concrete Work*

a. Concrete exposed to early drying must be kept dampened for at least a week after placing. Defective concrete resulting from any incorrect method of mixing or placing shall be removed at the expense of the Contractor and replaced.
b. Mix and place only when temperature is 40 degrees F or higher, and rising, unless permission obtained from Architect or Engineer. When surrounding air temperature is below 40 degrees F, concrete to have temperature of 60-90 degrees when placed. Maintain at temperature of at least 50 degrees for 72 hours after placing except for high-early cement when this time may be reduced to 24 hours. Method of heating materials and protecting concrete to be approved by the Architect or Engineer.

8. *Mixing Concrete*

Concrete, whether transit, Job or plant ready, must be mixed in an approved type of power operated mixer that will insure uniform distribution of material throughout, in a sufficient number of mixers to rapidly carry on work. Mix no less than 1½ minutes after all ingredients are in the mixer, including water.

9. *Concrete Tests*

Cylinders for compression tests shall be taken at each of four different locations in footing pour. Furnish three cylinders for each test; one for 7 day test, one for 28 day and one for spare.

Compression tests shall be made by a recognized testing laboratory and costs of the tests shall be paid for by the Contractor.

10. *Concrete Floors, Walks, Slabs and Drives*

All floors shall be trowel smooth; all walks and driveway shall be trowel smooth and broom slightly all wearing surfaces. All finishes shall be approved by the Architect or Engineer.

11. *Foundations*

a. Materials—All materials shall be of good quality and shall meet the requirements described in Division 9.
b. Footings—All footings shall rest on solid ground. No footings on tamped fill with dirt.
c. Foundation Walls—shall extend a minimum of 6" below frost line.

12. *Construction and Expansion Joints and Scored Joints*

a. Before beginning to pour concrete, the Contractor must secure special instructions from the Architect or Engineer relative to the proper location of the joints in the work. No joints shall be made in any concrete work except at such points as may be directed by the Architect or Engineer.
b. Expansion and control joints shall be provided in all concrete where indicated and as detailed.
c. Provide scored joints in all areas scheduled, or shown to have finished floors of concrete or cement. (Not required where soft finish floors are scheduled.)

13. Monolithic Finish

a. Where monolithic finish is called for on concrete floor surfaces, it shall be as hereinafter specified.

b. Monolithic finish shall be produced without a separate topping by striking off and screeding the concrete to a true surface at the required elevation. The concrete shall be tamped with special tools to force the aggregate away from the surface. While the concrete is still green but sufficiently hardened to bear a man's weight without deep imprint, it shall be wood floated to a true, even plane with no coarse aggregate visible. The concrete shall then be troweled to produce a smooth, impervious surface, free from trowel marks. No dry cement or mixture of dry cement and sand shall be sprinkled directly on the surface of the wearing course to absorb moisture or to stiffin the mix.

14. Concrete Floors on Ground

a. Concrete floors resting on the ground shall have monolithic finish without topping, as above described and specified.

b. Except where otherwise designated, the concrete floors on ground shall be 4" thick (total thickness.) Reinforcing is to be as specified under "Reinforcing Steel."

c. Provide steel troweled finish were scheduled and under all composition flooring.

d. Where quarry tile finish floors occur on concrete floors on the ground, depress the 4" slab one and one half inches (1½) and omit the troweling.

e. Provide moisture barrier under all first floor concrete slabs. Material shall be one layer 0.004" thick "Plasex" plastic sheet or approved equal.

1. General Notes

Reinforcing concrete construction is to be provided where indicated on the drawings or specified.

2. Material and Tests

All reinforcing bars are to be in accordance with the Standard Specifications for Steel Reinforcing Bars as adopted by the American Association for Testing Materials and shall be hard grade preformed bars of new billet steel or rerolled rail steel.

The number, size, spacing, position and bending of the various bars are to be as indicated on the drawings and schedules and no variation whatsoever is to be made without the written consent of the Architect or Engineer. Should the exact spacing or bending of the bars not be clear from the drawings, the Contractor must secure exact instructions from the Architect or Engineer before placing this work.

3. Placing

All reinforcing steel for all reinforced work must be accurately wired in position and properly spaced. Furnish and put in place for all reinforced concrete work, construction rods, spacers, ties and bar chairs in accordance with the current manual of the Concrete Reinforcing Steel Institute.

4. Shop Drawings

Furnish in triplicate for the Architect's or Engineer's approval, an accurate and complete set of shop drawings showing the steel reinforcement and all structural details of the reinforced concrete work throughout the building. All openings in reinforced floors must be carefully shown and framed and accurately dimensioned on the shop drawings.

5. Reinforcing Steel

Reinforcing steel shall be clean and free from rust, scale or coatings that will reduce bond.

6. Steel Welded Wire Mesh

6 x 6 No. 9 wire shall be used for all walks and all other concrete slabs.

1. General Notes

Furnish and put in place all masonry work as indicated on the drawings and herein described.

2. Mortar (All Proportions by Volume)

Cement mortar shall be composed of one part Portland Cement and three parts sand, to which may be added not more than 1/4 part hydrated lime or lime putty to the combined cement and sand.

3. Mortar Waterproofing

Waterproofed cement mortar shall be cement mortar made waterproof by using either a waterproofed Portland Cement of by adding to the mixture a waterproofing compound which shall be mixed in accordance with the manufacturer's directions.

4. Protection

All work and material must be protected from the weather and the Contractor shall be held responsible for damage to the same.

5. Parging and Jointing

The Contractor must see to it that shoved work is used exclusively in laying face brick and common brick and that each course is thoroughly grouted with cement mortar before the succeeding course is put on. The inner surface of all face brick or common brick shall be *fully parged and all head joints shall be* filled. Bed and head joints of hollow units shall be full over the uncored area.

6. Cleaning

After completion, all masonry, both inside and outside the building, shall be thoroughly cleaned down with a weak solution of rainwater and muriatic acid. Care must be taken to allow none of the acid to come in contact with materials other than the brick. The Contractor will be held responsible for other materials injured by acid.

7. Interior Walls and Partitions

a. Contractor's special attention is called to the fact that certain interior walls and partitions are to be of masonry and that most of the masonry work, including the inside face of exterior walls, will be exposed.
b. Careful workmanship, resulting in walls giving a finished appearance, will be required. Discolored or broken units, smeared surfaces, and joints of uneven thicknesses will not be tolerated.
c. Contractor's attention is called to the fact that modular units are specified, and when these are laid with 3/8" joints they should coordinate with the face bricks which are to be laid three courses to 8 inches.

8. Cutting and Fitting for Switches and Piping, Etc.

The Contractor shall do the necessary cutting and fitting of all masonry units around switch boxes, piping, fixtures, conduit, etc., to insure a neat and finished appearance.

9. Site Curing and Protection of Concrete Units

a. All concrete block masonry units, including both sand-gravel or stone aggregate blocks and insulating concrete block, shall be cured and protected at the site as hereinafter described.

b. All units shall be brought to the job site and stored thereon for not less than 28 days before they are incorporated in the construction. Such storage shall be arranged in a manner as to keep the blocks dry, either by tarpaulin or shed roof or combination of same, and blocks shall be stored off of the ground by means of planks and skids. Blocks stored in contact with the ground shall not be used. Storage shall provide for a reasonable amount of natural ventilation.

c. After the units have been incorporated in the walls and partitions of the building, the tops of the walls shall be protected against the entry of rain by means of tarpaulins, planks, or other suitable cover at all times when masons are not working. and such protection shall be continued until walls are under roof or coping.

d. The above requirements for curing and protection need not be applied to foundation walls below grade, tunnel walls, dwarf walls, and walls surrounding crawl spaces, where such walls are not exposed to view.

10. Masonry Wall Reinforcing

a. Furnish and put in place horizontal reinforcement for all masonry partitions and walls as hereinafter described.

b. The reinforcement shall consist of two continuous No. 9 steel wires laced together similar and equal to "Ninor" or eight No. 18 gauge longitudinal wires laced together similar and equal to "Super." It shall be of proper width for installation in the nominal 4", 6", and 8" concrete or insulating concrete block unit as necessary. In the case of 12" units, two widths of 4" reinforcing shall be used. Exterior walls shall have 12" widths.

c. Wall reinforcing shall be fully embedded in the horizontal mortar joints and shall be continuous. All joints shall be fully lapped and reinforcing shall be carried around corners by cutting and bending. Terminate reinforcing ½" shall *not* be bridged. The wall reinforcing as above described must be provided in every third bed joint (every 2'–0" vertically) of all interior and exterior masonry walls and partitions throughout the entire building. The first layer shall be placed at one course above finish floor line and succeeding layers every 2'–0" vertically thereafter.

11. Laying of Masonry

a. All work must be carried up with courses plumb, straight and true, with courses level from one end to the other.

b. Mason must level all walls for bearing by rod heights to receive joists, beams, columns, griders, etc., with blocking. Rods must be used at least every ten feet in length of bearing walls and at all piers, corners, plates, beams, etc. Mason will be required to clip bricks wherever necessary to gain the required heights. Mason must cooperate with other contractors in this work.

c. Face brick and exposed concrete masonry units are to be laid true to line and level, with horizontal and vertical joints of uniform thickness as designated. The courses shall be exactly poled to line with all windowsills and heads and wherever necessary. Vertical joints must be kept in alignment throughout and symmetrically placed in panels and piers. Where jointing is shown on the drawings, it must be followed exactly. The necessary cutting of block and units to execute the work shown on the drawings shall be done.

d. All exposed joints generally shall be concave joints which shall be accomplished with a tool suitable for the purpose. Joints in masonry below grade and in pipe tunnel and crawl spaces, above ceilings, behind cabinets, shelving, etc., may be struck smooth.

e. All hollow units on which the ends of steel or concrete beams or columns bear, shall be filled solid with concrete or mortar to a depth of two courses (16") down from the bearing course. The top course (8" high) of all units on which the ends of steel joists bear shall be either of solid units or filled solid with concrete or mortar.

12. Materials

a. *All masonry walls* not otherwise designated or otherwise hereinafter specified are to be built in mortar of a kind elsewhere specified of the following materials:

   1) Face brick shall be provided for the outside surfaces of all exterior walls where so indicated on the elevations. Face brick shall commence from one brick below finish grade line and extend up to top of wall as detailed.
   2) The backup courses directly under all floor and roof joists and beams shall be of solid concrete units or solid brick units for at least *8"* in vertical height.
   3) All exterior and interior foundation walls generally shall be concrete block except where poured concrete is called for on the drawings.

b. *Concrete blocks* shall be load-bearing concrete units and shall consist of 2 core units of nominal face from 4" by 12" to 8" by 16" either 4" or 8" thick, made with aggregate of gravel or crushed stone, and shall conform with standard specification of the ASTM No. C90 latest edition. Minimum ultimate crushing strength over the gross area shall be 700 pounds per square inch.

c. *Insulating concrete block* shall conform to the requirements hereinbefore set forth for concrete blocks, and in addition shall have insulating and sound-absorbing characteristics and shall, in general, weigh about one-half as much as the concrete block of corresponding size. All insulating concrete block shall be made of either an expanded clay or shale aggregate or a pumice aggregate. Pumice shall be uniform graded pumice aggregate in accordance with the standards of the Pumice Producers Association. Expanded clay or shale aggregate shall be in accordance with the standards and recommendations of the Expanded Clay and Shale Association of Allentown, Pa. All insulating concrete block must have a minimum crushing strength of 700 pounds per square inch measured over the gross area of the block and shall conform to the latest edition of ASTM Specification No. C-90, Grade D. Minimum base thickness shall be one and one quarter inch.

d. *Face Brick* shall conform to ASTM Specification C216-50 Modular Size, Grade MW, Type FBX. Color and texture as selected by Owner. Stack facing brick at site and avoid chipping. Face brick for exterior walls above grade as indicated. Estimates to be based on brick costing Sixty Dollars per thousand ($60.00 per M) delivered and stacked at the site. Adjustment will be made in contract price to agree with price of brick actually selected.

13. Insulation

This Contractor shall fill cores of all hollow lightweight Masonry Units and 2" cavity with water repellent Vermiculite Masonry Fill Insulation. Filling shall be done at window-sill line and before wall is capped. The water repellent Vermiculite shall be Lite Masonry Fill Insulation as manufactured by Lite Company or equal. "U" Factor of combined wall must equal .17.

14. Contraction Joints

Contraction Joints (C.J.) in concrete block walls shall be located where indicated on plans and where not indicated they shall be provided on interior surfaces of exposed block walls so that not more than 20 feet of block masonry wall occur between vertical contraction joints or intersecting walls. Contraction joints shall consist of vertical joint from bottom to top of the blockwork in the wall. After laying, the joint shall be raked out on the exposed face for a depth of 3-8 inches and tooled smooth and firm with a narrow pointing trowel. Do not cut wall mesh at contraction joints.

15. Rigid Insulation

The perimeter of the building shall have 1" Styrofoam extending 24" horizontally and 24" vertically as shown on the drawing.

16. Precautions

Do not lay masonry in freezing weather unless suitable means are provided to heat materials, protect work from cold and frost and insure that mortar will harden without freezing. No antifreeze ingredient shall be used. When outside air temperature is below 40 degrees F, the temperature of the masonry when laid shall be above 40 degrees F, and an air temperature 40 degrees F shall be maintained on both sides of the masonry for a period of at least 72 hours.

DIVISION 10

1. General Notes

a. Stone shall be red sandstone cut to fit in the same measurement as standard concrete block.
b. Stone shall be quarried from South Dakota.
c. Stone shall be even run in color, with all having the same curing time.
d. Stone shall be shipped to the job site in one shipment to assure the same texture, grain and color.
e. Stone shall not be chipped or cracked.
f. Stone shall have a 3/8" bond joint and all up the same as masonry.

1. General Notes

a. The underside of all concrete slabs on earth shall have been laid on milcore 0.004 inch thick.

b. Any portion of the foundation that is below earth or grade level shall be given a hot tar coating in the following manner.

  1) One hot tar coating shall be mopped or spread on and be given at least eight hours for setting time. The second coat shall be applied immediately thereafter.

c. This contractor shall be responsible for any leaks in the foundation that occur as a result of rain seepage.

1. General Notes

a. Provide all structural steel as shown on the drawings and herein described.

b. In general, the drawings will show the dimensions, weight, location and shape of all structural steel members, the various steel beams, girders, or members being represented by a heavy dot and dash line through the center of gravity of the member.

c. In general, no attempt will be made to show on the drawings any details of assembly such as connections, plates, bolts, anchors, rivets and the like. These are in all cases to be supplied by the structural steel fabricator and are to be in accord with the requirements herein specified.

d. Structural steel fabricator shall investigate availability of sections indicated on the plans before submitting proposal, and shall include any extra cost of substitute sections. Use of substitute sections of similar properties and dimensions will in general be permitted at no extra cost to the Owner.

2. Materials and Tests

a. All structural steel shall conform to the Standard Specifications for Structural Steel for Buildings, as adopted by the American Society for Testing Materials.

b. Tests, if required, will be undertaken by the Owner at no cost to the Contractor.

3. Fabrication

a. All structural details not otherwise shown or otherwise herein provided for shall be in accord with the standards indicated in the latest edition of the handbook of the American Institute of Steel Construction.

b. All connections must develop full strength of the member except where a lower stress is indicated on the drawings.

c. All truss joints, whether made in shop or field, shall be riveted and/or welded. In general, welded connections will be permitted in lieu of riveted connections providing they are of equivalent strength. Shop connections in general shall be riveted and/or welded, while field connections may be riveted and/or welded and/or bolted.

d. Connections generally shall be designed to come within the finished surfaces of the walls, floor, and ceilings, and no protruding connections shall be permitted without special permission of the Architect or Engineer.

e. Provide all drilling or punching of structural steel requested by the General Contractors in advance of fabrication, for attachment of anchors, bolts, fasteners, hangers, etc.

f. Provide welded anchors for structural members where indicated on drawings, and wherever else required to insure lateral stability of the structural shape. Space anchors not to exceed ten times minimum width of the structural members. Provide anchors wherever structural steel abuts against concrete construction on one side only.

4. Loose Lintels

a. Provide structural lintels as scheduled and shown; assemble and weld as indicated.

5. Erection

The contractor shall erect all structural steel in general accord with the Code of Standard Practice for Steel Buildings and Bridges, latest edition of the American Institute of Steel Construction. No special method of sequence of erection is required.

6; Shop Painting

All structural steel shall be shop painted one coat of red lead in oil.

7. Shop Drawings

a. The structural steel contractor must prepare and submit in triplicate to the Architect or Engineer for approval complete shop drawings, showing all details of framing, connections, bearing plates, anchors, and the like, and shall execute the work indicated only after the Architect's or Engineer's approval is obtained.
b. In the preparation shop drawings, the structural steel contractor shall check all dimensions and quantities shown on the drawings, and shall notify the Architect or Engineer at once of any discrepancies.
c. Should any necessary information as to dimensions, location or level be lacking on the drawings, the Architect or Engineer will, upon request, supply such information. In case of work set by others, the structural contractor shall supply full information as to portions of work delivered and the responsibility of this contractor shall not cease until such work is properly located in the structure.

1. General Notes

a. Furnish and put in place steel joist construction where indicated or scheduled.
b. Steel joist construction is designated on the drawings by the letters "S.J." followed by a number indicating size of joists required.

2. Handling and Storing

In shipping steel joists, in loading and unloading, and in handling at the site, due care shall be taken to avoid bending or distortion of the members. Steel joists, when delivered at the site, shall be placed upon skids or trestles which will properly support them and keep them from contact with soil or refuse until erected.

3. Fabrication

Steel joists shall be of the type noted on the structural drawings, conforming to the latest requirements of the Steel Joist Institute. Joists shall be as manufactured by Kalor Steel Company, Stram Steel Products or Brander Steel Company, or any member of the Steel Joist Institute. Joists manufactured by companies other than those specified above **will not be acceptable.**

4. Shop Painting

All Joists and accessories shall be given one (1) coat of rust inhibitive paint (standard with the manufacturer) at the factory.

5. Erection

a. Open truss-type steel joist construction shall be in accordance with the Standard Specifications for Steel Joists as adopted by the Steel Joist Institute. All Shortspan Series joists shall be not less than four (4) inch bearing on masonry walls and not less than two and one half (2½) inch bearing on steel supporting members. Longspan joists shall have six (6) inch bearing on masonry walls and four (4) inch bearing on steel supporting members.
b. Each steel joist shall be welded to the steel supporting beams. Where steel joists rest upon exterior or interior masonry walls, concrete beams or walls, they shall be anchored to same by means of one Government anchor to every other joist. Anchorage to steel beams shall be by welding equal to strength of two (2) three-fourths (¾) inch bolts for Longspans, two (2) welds, one (1) inch long for Shortspan Series.
c. The utmost care shall be taken in handling, storing and setting joists to avoid rust spots, scratches, chipped paint, etc. All such damage shall be touched up by the Contractor at the job. Touchup coat shall be brushed on.

6. Special Fittings and Bridging

a. Steel joists shall be bridged with rigid bridging as indicated and in accordance with the Specifications for Steel Joists as adopted by the Steel Joist Institute.
b. Steel joists parallel to walls shall have top and bottom chords anchored to the wall where each row of bridging occurs.
c. For all portions of the building throughout where open truss steel joist construction forms the ceiling of the area, furnish and put in place for each end of each joist, securely attached to the bottom chords of the joist, a furring extension rod or furring extension unit to properly support the furring, channels, or ceiling lath for plaster or acoustic tile.

d. *Special Provisions for Expansion*—Contractor's attention is called to the fact that not all joists are to be welded to supporting beams. In numerous places, joist ends will be bolted and/or clipped to supporting beams or wall supports or placed into special pockets in masonry walls in such a manner as to permit expansion. Contractor shall consult Framing Plans for locations and details.

e. Ceilings attached to underside of joists shall have 2" x 2" furring strips for insulation board ceilings. Refer to Room Finish Schedule for rooms to have insulation tile ceilings. *This contractor to furnish furring clips* for roof screeds and for ceiling screeds. Clips to be same or equal to Truscon Furring Clip No. P-3180. Two clips are required at each intersection of *roof screed and joist* and for *intersection of ceiling screed and joist.*

7. *Alternate Proposals*

This Contractor shall make provisions in the bid if alterations in the steel joists are needed for Alternate Proposals in Division 3. All changes shall be noted in submitted shop drawings.

8. *Shop Drawings*

Refer to Division 1, Paragraph 13, Page 1-4.

a. The steel joist contractor must prepare and submit in triplicate to the Architect or Engineer for approval complete shop drawings showing the location, size and length of all joists and all accessories connected with the steel joist construction.

b. In the preparation of shop drawings, the steel joist contractor shall check all dimensions and quantities shown in drawings and shall notify the Architect or Engineer at once of any discrepancies.

9. *Reinforcing Steel*

Refer to Division 8 of the Specifications.

1. General Notes

This work shall include all labor and materials required to install the carpeting system where designated on the room finish schedule, having to do with the carpeting.

2. Material

Carpeting shall be 100% worsted wool, with a face of at least 60 ounces of yarn per square yard. Carpet shall be tufted into a durable jute back and finished with a second back cover of jute. The weave shall be such that the pile will not flatten or mat during the normal traffic load. Carpet shall be recognized as commercial carpeting. The color shall be as selected by the Architect along with the weave and pattern. Samples shall be submitted to the Architect and Owner. The manufacturer shall be Conex Company or approved equal. The Contractor shall be allowed a minimum of $15.00 per square yard which shall be included in the base bid of the contract. Any amount over and above the amount of $15.00 per square yard will be in addition to the contract price and shall be paid for by the Owner.

The underlayment of the carpet shall be synthetic foamed rubber, or polyester foam rubber, a minimum of three-fourths (¾) of an inch thick. Underlayment shall be spliced only as necessary. Underlayment shall be installed at least twenty-four hours before the carpet is stretched in place.

3. Storage

Carpeting shall be ordered when the Contractor submits the pattern to the Architect for approval. The carpet shall not be stored on the job site, or within the building during construction.

4. Installation

Carpet shall be installed by the Contractor or his subcontractor who is an experienced carpet layer. The floor shall be thoroughly clean of dust, dirt, oil and other chemicals. The floor shall be completely dry. The carpeting shall not be installed until all workmen have completed their work in that particular area. This includes all testing of plumbing, heating, electrical, telephone and other systems. When the carpeting is completely installed the room shall be locked and the key turned over to the Architect.

5. Application

Carpet tack strip shall be placed with shot screws. The load shall be such that the seat will be flush with the tack strip. The shot screw shall not split or break the tack strip, nor shall the tack screws break the concrete. Where the carpet ends in open spaces, the metal tack strip with foldover edge shall be installed.

6. Guarantee

The installation and the carpet shall be guaranteed by the manufacturer for a period of five years. Such guarantee shall not include injury as a result of movement of mechanical equipment, and burns of color fading due to chemical action of cleaners.

1. General Notes

This work shall include all labor and materials required to install the entire roof deck system.

2. Material

The roof deck shall be Petrel Roof Deck or equal as manufactured by Louis Petrel Company. All substitutions must be approved by the Architect or Engineer before bidding. The insulation factor in the combination of built-up roof shall be 0.103 U factor.

3. Storage

Storage of the deck on the job site shall be on an elevated platform and the entire storage file covered with tarpaulin or waterproof paper.

4. Erection

Inspection of spacing of the supporting steel prior to the hoisting of the deck to the roof shall be made to insure the most accurate installation. All Petrel Roof Deck laid in one day should be made watertight prior to the completion of the day's work.

5. Application to Bar Joist

Clipped to bar joist and Petral Slabs shall be laid at right angles to the bar joist with the ends of the slabs bearing equally on all framing members.

6. Alternate on Steel Roof Deck (G-2)

a. An alternate proposal is desired, stating the amount to be added or deducted from the base bid, in the event that all areas specified in the base bid to have Petrel roof decks, except over Multi-Purpose Room, are provided with steel roof deck.
b. All roof decks affected by this Alternate will be as shown on Plan Sheet G-1.

TRANSLUCENT STRUCTURAL PANEL SPECIFICATIONS DIVISION 16

1. Materials

a. *Translucent Facing*—shall be fiberglass reinforced plastic sheets that are uniform (1) in color to prevent a splotchy appearance, (2) in thickness to assure proper bonding strengths and impact resistance, and (3) in strength.
   1) Exterior face shall have a special protective surface for maximum resistance to erosion and weather. This surfacing shall be applied and cured simultaneously with the basic sheet in order to produce a chemical bond. Post-applied sprays will not be acceptable.
   2) The faces shall be completely free of ridges and wrinkles which prevent proper surface contact in bonding.
   3) Clusters of air bubbles and pin holes which collect moisture and dirt will not be acceptable.

b. *Laminate Adhesive*—waterproof resin type adhesive especially compounded for use in laminating plastic sheets to aluminum grid members, capable of withstanding impact and temperature shock normally encountered in exterior construction.

c. *Grid Core*—shall be 6063-T5 aluminum with provisions for mechanical interlocking of members.

d. *Windows*—General
   1) All windows to be Core Standard Clamp—tite series, as manufactured by the Core Corporation, Clele, New Hampshire, designed specially for inclusion in the Core Panel Unit system.
   2) *Materials*—all window members shall be of 6063-T5 extruded aluminum mill finish and designed for exterior putty glazing. The overall depth of the window at the ventilator shall not be less than 1½" with 7/16" glazing legs.
   3) *Construction*—all window frames and vent components are joined together by a mechanical swedge to provide a permanent, rigid connection. All joints exposed to weather shall be sealed with a continuous elastic non-hardening compound to provide a permanent watertight connection.
   4) *Ventilators*—vents shall be of the project-out type with mitered frames allowing approximately 60 degree opening and heavy duty friction shoes shall be fully adjustable. All hardware to be white bronze. Cam locking handles shall be designed for hand operation.
   5) *Screens*—screens shall be made of rolled section aluminum frames, mitered and joined by consealed aluminum corner angles. Screen cloth shall be 18/16 aluminum mesh and held in place by vinyl plastic splines. Screens to have wicket for access to handle.
   6) *Glazing*—is under Glazing, Division 29.
   7) Ventilators shall be weatherstripped around their entire perimeter with vinyl weatherstripping.
   8) After fabrication, windows and mullions shall be chemically cleaned and etched, followed by a dip coat of methacrylate lacquer.
   9) Windows shall be set plumb and true in openings, securely, anchored and held in place during construction. Allow for expansion between and adjacent construction, including mullions; shall be sealed with gaskets or sealant. Sealant to be furnished and applied by the erector. Aluminum surfaces which come in contact with dissimilar materials shall be coated with alkal-resistant paint by the erector. Care shall be taken that lime mortar, plaster or any other corrosive substance shall be promptly removed from exposed aluminum surface. Ventilators shall be carefully adjusted and hardware attached before glazing.

e. All panels and insulation to be standard white apaque.

2. Insulation

The fiberglass pack insulation shall be of sufficient density to create overall light transmission of 20% and "U" value of 0.25, specially processed to insure non-sagging light stability and color as per sample. Manufacturer must certify that the insulated panel has been in satisfactory use for at least 3 years.

3. *Bonding Strength*

The general Contractor shall erect Core Panels in strict accordance with standard installation details or approved shop drawings as supplied by The Core Company. Fastening and sealing shall be in strict accordance with the manufacturer's recommendations. Core Panels shall be carefully stored under cover by the Contractor before erection; all aluminum shall be clean before sealants are applied. Sealants are to be clean and free from foreign matter prior to application.

Isolate aluminum ferrous metals to avoid damage by electrolysis. Do all finishing work including caulking as required.

After the other trades have completed work on the adjacent material, carefully inspect panel installation and make adjustments necessary to insure efficient installation and weathertight condition. When requested by Architect or Engineer, this contractor will conduct hose test to assure weathertight installation.

4. *Battens and Perimeter Closures*

a. Extruded 6063-T5 aluminum members
b. All battens and perimeter closures to be supplied with stainless steel screws (excluding final fasteners to building)
c. Receiving channels for self-tapping stainless steel screws to be continuous the length of each member

5. *Elastic Sealing Tape*

The elastic sealing tape will be factory recommended extruded butyl compound with string center pre-applied to closures at the factory.

6. *Construction*

a. Panels shall be sandwich panels of flat fiberglass sheets bonded to grid structure of mechanically interlocked aluminum I-beams.
    1) Panels shall not rely upon rolled or banded edges for support of skin edges.
b. Grid core shall be mechanically interlocked to assure even bonding surface contact with face materials and resistance to torsional stress.
c. Panels to be laminated under a controlled eight hour heat and pressure process.
d. Grid pattern to be width required to fit 4'0" opening (approximately 8") by 20" long and each grid to span half of two grids on either side.
e. The adhesive bonding line shall be straight, cover the entire width of the I-beam and have a neat sharp edge.
f. To assure complete bonding contact, no white spots shall appear at the intersection of mutins and mullions in grid structure.
g. Unit shall be 2¾ inches thick.

7. *Protection and Cleaning*

The general contractor shall be respnsible for protecting the windows during construction and for cleaning at the completion of the building.

8. *Caulking*

The caulking contractor shall caulk all joints between window frame and masonry with suitable materials.

9. *Shop Drawings*

The insulating panel contractor shall supply the Architect or Engineer with shop drawings in triplicate showing all details of manufacturer number and location. The manufacturer of the panel is not to proceed until shop drawings are approved.

1. General Notes

This contractor shall have fabricated all ornamental iron work as required to complete the project.

a. Material shall be unused cold rolled iron, without rust cavities.
b. Installation of the finished fabricated product shall be as designed on the plans in the detail section. Care must be taken to prevent cracking of the stair treads during installation.
c. All ornamental iron shall be given two coats of enamal after cleansing of all dirt and grit.

MISCELLANEOUS METAL AND METAL SPECIALTIES — DIVISION 18

1. General Notes

The general contractor shall furnish and install miscellaneous metal indicated on the drawings including railings, scrapers, etc.

2. Foot Scrapers (Exterior)

Furnish and install foot scrapers at each main entrance as shown on Plot Plan and detail drawings. Scrapers shall be fabricated of steel plate with one-fourth inch (¼) scraper blades ground to a one-eighth inch (1/8) edge. Scrapers and installation shall follow detailed drawings.

3. Metal Letters

a. Letters are to be as follows:
   (THE NAME OF THE PROJECT).
b. Cast Aluminum, Black baked enameled finish.
c. 6" high x 1/2" deep with 1" stroke.
d. Provide studs for type F–flush mounting on masonry.
e. Provide full size template for mounting with proper letter spacing.
f. Provide shop drawings in triplicate for Architect's or Engineer's approval.

4. Metal Partitions

Furnish and install, in the girls and boys shower and toilet room, metal showers and toilet screens as indicated on the floor plan drawings. Partitions and screens should be installed with all necessary hardware, braces, and fittings as required for a complete installation. Include paper holder No. R-200 for each water closet.

The entire partition and screen installation shall be full flush type, overhead braced partitions and screens as manufacture by Metalic Partitioning Corp. or approved equal. The overhead brace of the shower stall is to be used as a curtain rod.

Shop drawings in triplicate must be submitted to the Architect or Engineer before final approval and installation may begin.

Colors of partitions and screens in the shower room shall be single color for partitions, screens, and compartment doors. One color scheme will be selected by the Architect or Engineer (Manufacturer's standard colors) for use throughout the building.

5. Installation

Installation shall be in accordance with the approved manufacturer's shop drawings and all anchors at floors and walls shall be rigid and secure. Particular attention shall be given to the anchorage into finished materials so that anchor holes are clean and unbroken.

6. Toilet Room Mirrors

Furnish and install one 16" x 24" mirror over each lavatory except in rooms 109 and 110.

Each mirror shall be polished plate glass with concealed mounting and shall be similar and equal to Glasco.

7. Toilet Paper Holders

Furnish and install one (1) Mode Roll Dispenser No. 6 or equal for each water closet except the water closets within metal partitions which shall be supplied by the metal partition manufacturer.

8. Paper Towel Holders

Install Mode Public Service No. 10 Paper Holders or equal in the following Rooms:

Room Nos. 101 and 122–2 each
Room Nos. 124, 110 and 109–1 each

9. Medicine Cabinets

Provide recessed medicine cabinets in rooms 109 and 110 with the special child-proof latch. Cabinets shall be Medic No. 103 or approved equal.

The General Contractor is to supply 1/4" steel plate lintel with minimum 4" bearing at each side of recess opening.

10. Pass Thru–Dirty Dish Window

Provide Grere 1¼" flat slat rolling shutter galvanized steel 22 U.S. gauge with grey enamel primer. Doors shall be between jambs, hand crank operated. Furnish and install as shown on plans.

a. All steel door frames and hollow metal sidelights shall be as manufactured and furnished by Pass Products Corporation or approved equal.
b. Frames to be minimum 16 gauge hot rolled steel to manufacturer's standard contour dimensions and to frame widths indicated on drawings. Frame to be mitered at joints and arc-welded internally, and to be provided with holes in strike face of stop for standard floor silencers. Furnish rubber silencers for installation in field. Frames to be provided with hinges specified in hardware section. Frames to be cut for strikes and reinforcements welded, and with plastic cover welded to strike and hinge reinforcement.
c. Frame to be given manufacturer's standard shop coat finish.
d. Provide shop drawings in triplicate for Architect's or Engineer's approval.

1. General

a. Provide sheet metal work as herein described and/or shown on the drawings.
b. Flashing for plumber's outlets are *not* included in this Division.

2. Materials

All sheet metal not otherwise designated shall be galvanized sheet metal gauge as indicated on drawings.

3. Counterflashing Over Composition Roofing

a. Counterflash bituminous base flashing at all vertical surfaces, walls, curbs, etc., throughout.
b. Take measurements at building and fabricate all flashing. Miter lock and solder all internal and external corners, lap all other joints 6" and bed in roofers plastic cement, but do not lock, solder or rivet. Lap joints are to be spaced not more than eight feet apart.
c. In general, counterflashing against walls will be two-member flashing as detailed.
d. Turn up inside edge and hem outer edge of flashing as detailed.

4. Counterflashing for Roof Structures

a. Sheet metal contractor shall counterflash all roof ventilator curbs and all roof curbs generally unless otherwise detailed.

5. Mastic Sealing

a. Sheet metal contractor shall provide continuous mastic seal at certain critical locations in order to prevent the entry of water into the building. The locations requiring continuous mastic seal are indicated and detailed on the drawings and occur strictly in connection with items required as a part of the sheet metal work.
b. The following is a schedule of locations requiring continuous mastic seal.
   1) Between top of counterflashing and underside of roof ventilators
   2) Meeting of roof and vertical wall

6. Expansion Joints

Provide all sheet metal items required for all expansion joints as detailed. This includes all copper and aluminum sheet and all steel sheet up to and including 16 gauge. (Structural angles, and steel sheet heavier than 16 gauge are specified under other divisions.)

7. Galvanized Sheet Metal Gravel Stop

a. Furnish and install around perimeter of flat roof of building as detailed on drawings. See drawings for details.
b. Expansion joints shall be provided and all connections shall be sealed to prevent passage of moisture.
c. All corners shall be one-piece mitered and soldered.

8. Galvanized Sheet Metal Hoods

a. Provide and install dishwasher hood and range hood as shown on drawings.
b. Sheet metal shall be gauge indicated on drawings.
c. All corners shall be rounded and connection shall be soldered and rough edges ground off to provide smooth finished product.
d. Submit shop drawings of range hood and dishwasher hood in triplicate for Architect's and Engineer's approval.

1. General Notes

a. The work required under this heading shall include all labor, material, equipment and services necessary for and reasonably incidental to the proper furnishing and installing of all roofing for all roof surfaces.
b. The Contractor shall examine all roof surfaces on which or against which his work is to be applied for defects which he would consider detrimental to the installation of his materials and see that all defects are corrected; otherwise this Contractor at his expense shall replace all roofing materials which may have to be removed to correct these defects. The laying of materials shall be considered the acceptance by him of all surfaces.

2. Materials

a. All roofing materials shall bear the manufacturer's label on sealed package, and shall be the product of a nationally recognized firm with at least ten years in the industry.
b. Asphalt shall be equal to Byron 190 degree steep roofing asphalt.
c. Asphalt felt shall be equal to Bryon saturated roofing felt weighing not less than 14 pounds to the hundred square feet.
d. Coal tar pitch for roofing shall be Type A with a softening point of 140 degrees F to 155 degrees F and shall comply with ASTM Specification D-450-41.
e. Coal tar pitch saturated felt shall be composed of roofing felt impregnated with coal tar and coal tar pitch and shall weigh not less than 14 pounds to the hundred square feet and shall comply with ASTM Specification D-227-47.
f. Plastic cement shall be Type 11 composed of an organic filler, a solvent and a bituminuous binder, shall comply with Federal Specification SS-C-153.
g. Surface roofing material specified for roofing flashing shall be composed of asbestos roofing felt, one side of which shall be surface with mineral granules and weighing not less than 83 pounds per 100 square feet, and shall comply with Federal Specification SS-R-511.
h. Gravel used shall comply with Specifications for concrete gravel except it shall pass a 3/4 inch screen and all be retained on 1/4 inch screen.

3. Roof Deck Construction

a. Contractor's attention is called to the fact that the roof deck is as specified elsewhere.

4. Application

a. The application of roofing to the deck shall be regulated by manufacturer's recommendations. All recommendations shall be followed unless this Contractor has the Architect's or Engineer's written approval for variation.
b. Eave edges shall be checked to allow for release of air pressure under roofing membrane and care shall be taken in the application of the roofing so that pressure release will be operative at edges.
c. Channel-mop the entire surface in one direction only with continuous moppings of *steep asphalt* (high melting point asphalt) and embed therein one layer of 15 pound asphalt-saturated roofing felt, *lapping* 8" and mopping the lapping so that the entire surface of the roof deck is covered with felt. The channel moppings shall be continuous and extend from eave edge to eave edge so as to facilitate the release of air pressure at eave edges. The channels between moppings of steep asphalt shall be not less than 3" nor more than 6" wide.
d. After the first layer of felt is applied to the deck by channel mopping with steep asphalt, apply over the entire surface a heavy mopping of *coal tar pitch* and embed four layers of 36" wide tarred felts, lapping each sheet or layer 27½" over the preceding one, mopping felts in solidly so that in no place shall felt touch felt. Broom in felts immediately behind

mop to obtain solid bond and to eliminate void pockets between felts. Felts shall cover all cant strips but shall be cut off evenly at vertical wall lines. At open eaves or roof edges, extend the first two plies of felt 12" beyond the roof edge and cement these projecting felts back over the top layers forming a continuous "envelope" to contain the pitch. Use not less than 30 pounds of pitch per 100 square feet for each mopping.

e. After the four plies have been mopped in and enveloped at edges as above described, coat the entire surface with a heavy application (80 pounds per 100 square feet) of pitch poured from a dipper, and while hot embed not less than 400 pounds of gravel per 100 square feet over the entire surface. Not less than 200 pounds of pitch shall be used per hundred square feet for the embedment of the gravel and the mopping of the four plies of felt.

5. *Additional Requirements*

a. Pitch and asphalt shall not be heated above 400 degrees F and in no case higher than recommended by the manufacturer. It shall be periodically tested for temperature with thermometer.
b. All roof deck surfaces shall be smooth and firm, and shall be broomed clean of dirt and loose material and shall be thoroughly dry, whether it be a prefabricated deck material or a cast-in-place material, or over insulation which is not dry.

6. *Flashing*

a. Roofing felts and built-up flashings shall be applied at intersections of roof with all vertical masonry surfaces and at all curbs and roof edges.
b. Before applying flashing materials, the flashing grooves shall be thoroughly cleaned of mortar or other foreign materials.
c. All plies of the regular roofing felts and bitumen are to be extended up to the top of the cant strip against all walls, curbs, expansion joints, etc.
d. After the regular roofing has been carried up to the top of the cant strip, mop with steep asphalt and cant strip and the vertical surface, and mop on a tarred felt strip starting four (4) inches above cant strip and extending down over cant strip and out onto the roofing four (4) inches. On top of this apply a second strip of felt with mopping of hot steep asphalt, extending the felt onto the roof deck one (1) inch beyond the preceding strip. Mop on additional strips in a similar fashion until a four-ply reinforcement has been obtained.
e. Nail the top ply of flashing reinforcement on a center line of cant with one and one-half (1½) inch barbed roofing nails through flat tin disks on ten (10) inch centers.
f. Mop with hot steep roofing asphalt a strip of eighty-five (85) pound mineral surface roofing felt of sufficient width to extend from a line 4" above the top of the cant strip to the surface of the deck. Nail (into mortar joints) with one and one-half (1½) inch barbed roofing nails through flat tin disks at upper edge approximately ten (10) inch centers.
g. Tarred felt or mineral surface roofing felts are not to exceed ten feet three inches (10' 3") in length. Ends of connecting strips are to overlap three (3) inches and are to break at least twenty-four (24) inches with end laps of underlying strip. End laps shall be thoroughly mopped with steep asphalt.
h. All plies of felt shall be firmly pressed into moppings of steep asphalt so that there will be no wrinkles in the finished surfaces.
i. At roof edges where gravel stop is shown provide 3 additional plies of roofing on top of the regular roofing, feathering out onto the flat surface. Cover with the pour coat of pitch and gravel as specified under "Application on Roofing."
j. See details for roof and cornice flashings.
k. Copper counter flashing will be provided and installed as elsewhere specified. Gravel stops shall be mopped in as detailed on drawings.
l. Where vent and exhaust pipes and other obstructions extend through roof surfaces, cooperate with the Sheet Metal Workers and the Plumber and make all openings watertight.

7. *Insulation*

Provide and install one layer of rigid insulation board of 1/2" thickness for low roof area for installation with Metal Decking as specified elsewhere.

8. *Guarantee*

a. The Contractor shall provide to the Owner a written guarantee that all roofing is in strict accord with these specifications, in no case less than the roofing manufacturer's recommendations for a twenty (20) year bonded roof. This guarantee shall further agree to repair and replace any portion of the roofing, including flashing found defective or injured by water leakage, within a period of two (2) years from application of the roof, unless it be proved that such leakage or defect is due solely to abuse.

b. Roofers surety bond will not be required.

CAULKING AND WEATHERSTRIPPING DIVISION 22

1. General Notes

The building shall be carefully caulked and pointed as hereinafter specified and any points showing penetration of moisture shall be rendered tight at Contractor's expense and under the direction of the Architect or Engineer. Below grade caulking of vertical joints between existing and new addition shall be done before waterproofing. Seal joints around frames, sills of doors, windows, and other openings in exterior masonry walls and other joints or spaces noted to be caulked or pointed with mastic. All exterior thresholds shall be set in caulking compound. Seal joints between masonry and metal material. All caulking compound shall be Silicone construction sealant as manufactured by Silicone Sealant Company of Uresa, Illinois, unless otherwise specified. Caulking shall be delivered to site in unbroken containers bearing manufacturer's labels.

2. Application

After all cleaning operations on exterior building are completed, all areas to be caulked shall be cleaned of all foreign materials; then clean the surface of all areas, where Silicone Construction sealant is to be applied, with Silicone surface primer to insure good adhesion.

Joints that are wide and deep, such as expansion joints, shall have Etha foam rod stock of the size needed to fill the recesses before installation of Silicone Construction Sealant. Application shall be as recommended by the manufacturer.

3. Thresholds

Provide and install Valktite threshold or approved equal at all exterior doors.

4. Weather-stripping

All exterior doors and windows shall be weather stripped of approved material and installed following the manufacturer's recommendations.

1. General Notes

Furnish all material and perform all work specified in this Division. Woodwork not included in Division 24, Millwork, is to be included in this Division.

2. Material

a. The lumber grades hereinafter mentioned shall, except where otherwise specified, be those promulgated in the latest edition of the Standard Grading and Dressing Rules of the West Lumbermen's Association.
b. Wood for carpentry work generally shall be Construction Grade Douglas Fir for structural member. Hemlock for nailers.
c. Provide all rough hardware required for execution of the work such as nails, screws, bolts, anchors, etc. Thoroughly nail all framing in accordance with standard practice and in such a manner as to secure rigidity.
d. All exterior plywood shall be constructed with waterproof glue.
e. All Batt Insulation shall be 6" fiberglass. Insulate lower roof but above ceiling if steel deck alternate is accepted.
f. Supports and backing for cement asbestos board facing and application of same.
g. All shelving in storage rooms.

3. Cement-Asbestos Board Facing

a. Provide, at cornice locations, panel facing as indicated and detailed on the Plans and herein described.
b. Facing material shall consist of 3/8" thick flat cement-asbestos-board similar and equal to "Besto" as manufactured by Cementbest or "Cemboard" as manufactured by the Cembod Mfg. Company. The facing shall be applied to wood backing ar. nailing strips with rustproof nails, either galvanized or cadmium-plated steel or of nonrusting metal. Nails must be of the drive-screw type. Nailing shall be spaced approximately 8" on center. Holes for nails must be drilled. Contractor may, at his option, use brass screws in place of nails.
c. Provide solid wood backing behind all cement-asbestor-board facing consisting of 3/4" wood sheathing as detailed.
d. Provide, between the solid wood backing and the cement-asbestos-board facing, a continuous layer of 15 pound asphalt-saturated felt.
e. All cement-asbestos-board facing shall be primed on the back before erection with one coat of boiled linseed oil as elsewhere specified under "Painting Division." Exposed portions shall be painted after erection as elsewhere specified under Painting Division.
f. Furnish and put in place, behind each vertical joint in cement-asbestos facing a 4" wide strip of flashing material. This material shall be fastened to the wood backing and be placed to top of the 15 pound felt specified above. It shall be the same material as specified under "Flashing for Structural Steel" described under the Division entitled Miscellaneous General Contract Items.

4. Wood Plaster Grounds

a. All plaster grounds which are indicated on the contract drawings and details, and which are to be of wood, shall be furnished and put in place by the General Contractor under the Carpentry Division.

5. Shelving

a. Furnish and install all storage shelving as indicated on Plan.

6. *Alternate*

a. This Contractor shall provide and install all insulation hereinafter described or hereinafter scheduled as a part of steel deck alternate G-1.

b. All insulation shall be nominal 6" thick fiberglass blanket type or batt type with vapor barrior on inside.

c. Insulation shall be fastened together and laid over suspended ceiling where ceiling is suspended and laid tight and fastened together on ceiling applied directly to joist. All areas where batts or blankets cannot be fastened together shall be filled to batt depth with loose insulation.

d. Insulation as above described shall be required in the following locations.
   1) For the entire building with the exception of Multi-purpose room.

7. *Dimensional Lumber*

a. Joists of wood shall be as indicated on plans. Joists shall be installed with steel joist mounts or, if nailed, shall have a minimum of three No. 10 nails on each side where joist is nailed to sill box. Joist shall be Yellow pine with straight grain and free from knots. Joists must be high quality grade "A". Crest of Joist to be up. Joist be mounted square and level.

b. Studs shall be sized as indicated on plans. Studs shall be installed with two No. 8 nails on each side, top and bottom, where mounted on plates. Studs shall be of high quality, Grade "A", straight grain, free of knots. Studs must be mounted plumb and square.

c. Sheathing shall be 3/4" plywood manufactured specifically for that purpose. Sheathing shall be attached to studs, joist or rafters by nailing with No. 10 Common nails. Each nail driven with head flush with surface of plywood. Plywood shall be smooth and free from knots or voids on the side materials are to be attached.

1. Work Included

a. The work to be done under this heading includes the furnishing of all labor, materials, equipment, and services necessary for the proper completion of all exterior and interior wood finish, wood trim and wood cabinets work as shown on drawings and as specified.

b. Measurements for all millwork shall be checked and verified at building before fabrication of the various items called for to be fitted to special dimensions.

2. Materials

a. All exterior wood finish items shown or marked to be of wood, shall be Clear Western or Ponderosa White Pine.

b. Interior wood finish shall be as follows:

 1) All wood finish and all items shown on drawings to be of wood, shall be as detailed, scheduled or specified. Wood finish to be of kiln-dried stock, free from machine marks and rough spots, and hand sanded smooth throughout. All wood interior finish of every description shall be hardwood except storage room shelving.

c. All finish lumber and millwork must be protected from the weather while in transit; when delivered at the site it shall be placed immediately under cover or otherwise well protected from the weather. Such material shall not be delivered to the site until it is needed and shall not be stored within structure during process of Drywall finishing nor until the Drywall finish is reasonably dry and humidity conditions are suitable.

d. All finish millwork not elsewhere specified shall be Kiln dried, Plain Red Oak free from warp, shrinkage and discoloring pitch; selected birch suitable for natural or bleached finish and entirely free from marking that will show through finish, or select grade white oak.

3. Wood Doors

a. *Interior Doors (General)*–All interior wood doors species shall be indicated on door schedule. See Plans. All wood doors not otherwise indicated or scheduled shall be flush veneer doors 1-1/8" thick. Insofar as compatible with the requirements herein noted, stock pattern doors of reputable manufacture will be acceptable. Doors shall be either five-ply solid staved or framed core, or three-ply facing both sides over egg crate core. Doors shall have hardwood edge strips on sides. Doors shall be resin bonded by hot plate process. If hollow core doors are furnished, they shall be provided with solid lock blocks of ample size. Doors shall be provided with opening for glazing with hardwood stops where glazing is indicated or scheduled.

b. *Special Interior Doors*–Provide special interior doors where special sizes or patterns are indicated or scheduled. Fabricated to Architect's or Engineer's detail or to a detail approved by Architect or Engineer.

c. ***Exterior Doors***–All exterior doors shall be 1¾" thick. Exterior sash doors are to be solid mortised doors to detail. Exterior slab doors generally are to be solid waterproof resin bonded doors, or mortise frame doors covered with waterproof resin bonded plywood both sides. Double doors are to have astragals to detail at meeting stile.

4. Wood Cabinets

a. In general, all cabinet work indicated on the drawings is mill-made wood cabinets.

b. Cabinets shall be constructed of Oak or Birch 3/4" Veneer Plywood as indicated on the drawings, with concealed shelving of 5/8" Plywood with Oak or Birch edging where indicated. Counter tops shall be installed as specified elsewhere. Wood trim shall be Plain Oak or Birch. Cabinet doors shall be Oak or Birch 3/4" Veneer Plywood with square lipped edges.

5. *Shelving*

Shelving shall be made from 5/8" Plywood and shall be faced with a 3/8" x 1-1/8" Pine Molding on the front of the shelf and furnished widths indicated on the Plans.

6. *Installation*

a. Provide shop drawings in triplicate for Architect's or Engineer's approval of all cabinet work.
b. All miscellaneous items of finish and trim not especially mentioned in these specifications but shown or called for on drawings shall be installed.
c. All cabinet hardware shall be furnished and installed by this Contractor.
d. All door catches, hinges, and pulls must be approved by the Architect or Engineer.

7. *Competence*

The Architect or Engineer reserves the right to approve the millwork manufacturers selected to do all the millwork. The approved millworker must have a reputation for doing satisfactory work on time, and have recently successfully completed comparable work.

8. *Alternate G-4:*

The borrow lights as detailed on the drawing are not to be included in the base bid but to be figured as an alternate.

9. *Wood Windows*

a. This Contractor shall furnish and install wood windows as specified herein and at locations designated on Plan.
b. Windows shall be as manufactured by Woodcut Co., Inc., or Pinex and shall meet the following standards:
    1) Windows shall be size indicated on Plan.
    2) All windows shall be furnished with removable storm sash.
    3) All ventilating windows shall be furnished with removable extra strength aluminum frame. Aluminum Screen Cloth, 18 x 14 mesh, gun metal finish, lock applied hinges, strike plate, screws furnished.
    4) All ventilating windows shall be awning type windows and shall be Lock operated.
c. The insulating of windows shall be done in such a manner that all cracks and voids are filled with insulation and shall be inspected by Architect or Engineer before installing finish trim.

METAL LATH PLASTER, DRY WALL AND GYPSUM

DIVISION 25

1. General Notes

Where metal lath and plaster are indicated on drawings, substitutes will not be permitted. The metal lath and plaster shall be installed by such persons having a minimum of ten years experience. All plaster shall be plumb and level without defects. Such plaster shall consist of two coats. The first coat shall be 3/4" brown coat and shall not be applied in temperature less than 65 degrees F. The temperature shall not be allowed to drop below 65 degrees F for a period of seven days or until it has been determined by the Architect or Engineer that the plaster is sufficiently dry, to prevent cracking and freezing. When the brown coat is completely dry, the white cement finish coat of plaster shall be applied. The drying time for the finish coat shall be a minimum of seven days at a temperature of 65 degrees F.

Metal lath shall be expanded metal with stamped ribs. Metal lath shall be nailed to studs a minimum of eight inches on center, stretched tight prior or at nailing time.

2. Gypsum

Where gypsum board is indicated on the plans, the following shall be installed. All slabs shall be 3/4" thick, minimum width 48". Length shall be as required. The Contractor shall install gypsum board in such a manner as to eliminate as many joints as possible.

Where gypsum board is laid on masonry surface, masonry shall be smooth, level and plumb. Gypsum board shall be mounted with mastic applied at minimum spacing of four inches on center. Where gypsum board is mounted on wood furring strips, gypsum shall be nailed with wedge type nails.

3. Finishing

Each joint shall be given one thin cover of Joint cement. Then the fiberglass tape shall be applied. The final coat of mastic shall be finished with sandpaper and all joints are to be smooth. Upon completion of finish coat, the walls shall be ready for painting.

1. General

Where Special Coating is indicated on the drawings, it shall be "Polyester Coating" as specified herein. The work specified in this section shall be performed by an applicator licensed by the manufacturer of the material. He shall furnish to the Contracting Officer evidence that he has been a licensed applicator for the manufacturer of this material for a minimum of two years and list at least three applications of the material similar to this project.

2. Work Included

Provide and apply as specified, in Boys and Girls toilet rooms to height of 4' 0"; in Boys shower room to ceilings; in girls shower and dressing room to ceilings; in toilet numbers 110, 109, and 124 to ceiling; and in Kitchen to a height of 5' 4".

3. Materials

a. *Liquid* for polyester coatings shall consist of polyester Resin, 100 percent solids formulated by a coating manufacturer for use as a coating. The use of the general purpose resins will not be acceptable.
   1) The liquid shall be a thermal-setting plastic, with non-organic pigments milled in 100 percent reactive polyester and shall not contain evaporating type solvents.

b. *Hardening Agent* used to produce the cure shall be as recommended by the manufacturer, shall be non-toxic and shall be such that a chemical bond can be achieved between coats if succeeding coats are applied to fully cured coats within a two week period.

c. The *Materials* shall cure through chemical reaction into a 100 percent solid coating being completely non-porous, and exhibit no air-inhibited qualities and shall not support bacteria growth.

d. *Fiberglass Roping* shall be as specified by the coating manufacturer and tested to be compatible with the coating being used.

e. *Block Filler* shall be polyester block filler as specified by the coating manufacturer. No cementitious fillers will be acceptable.

4. Quality Standards

a. *Fire Rating*—Material shall have a fire rating of not more than the following as determined on the basis of tests conducted in accordance with ASTM Standard No. E84 and rated in accordance with the National Building Code of the National Board of Fire Underwriters:

Flame Spread	25
Fuel Contribution	Negligible
Smoke Contribution	25

b. *Abrasion Resistance*—Twenty thousand (20,000) cycles, no breakdown of coating. Materials, undergo complete cycle in Pangborn Sandblasting Machine without removing coating.

c. *Hardness*

Sward	46
Barcol	35-45
Compressive Strength	20, 715 psi

d. *Chemical Resistance*

Bulphuric Acid 30 Percent	Good
Sulphuric Acid 3 Percent	Excellent
Glacial Acetic	Fair
Five Percent Acetic	Good
Petroleum	Excellent
Benzol	Excellent
Turpentine	Excellent
Alcohol	Excellent
Ketones	Excellent
Salt Solution	Excellent
Lipstick, Ink, Milk, Iodine, Crayons, Mustard	Excellent

5. *Testing*

a. *Materials* delivered to job shall bear an Underwriter's Label indicating compliance with the test requirements herein, or if manufacturer does not have an underwriter's test and inspection service, the material delivered to job will be tested, at the Contractor's expense. Three such tests will be required during the course of the work. The materials shall be delivered and stored on the project site in sealed and labeled containers sufficiently in advance to allow 30 days for testing. Samples for testing will be selected by the Architect or Engineer.

6. *Samples*

a. Color samples shall be provided and approved before any of coating and priming procedures are started.
b. Contractor shall be required to supply samples of material in accordance with the specifications. Samples shall be applied vertically under supervision. Finished samples must be of prescribed mil thickness guage, and, after curing shall exhibit no inhibiting qualities in accordance with the following test. The cured sample shall be rubbed with the finger dipped in Acetone until all Acetone evaporates. Tackiness or pigment bleeding shall indicate an inhibited cure.

7. *Preparation of Surfaces*

a. *General*–The work described under this section shall be done prior to painting, decorating, and floor application. Quarry tile floor and base, where they occur, shall be finished before application of polyester coating.
b. *Concrete Masonry*–Work to receive polyester coating shall be properly cured and with a moisture content not to exceed ten percent. Excess mortar shall be moved and surface thoroughly cleaned. Any voids shall be filled with polyester block filler, prior to first coating.

8. *Application of Polyester Coating*

a. *General*–Finish application shall be not less than 30 mils on concrete masonry surfaces and 20 mils on plaster surfaces.

All voids in concrete masonry surfaces shall be completely filled as specified hereinbefore so that no pinholes remain after final coating.

b. *Operation No. 1*—Apply 6 mil coat of polyester coating.

c. *Operation No. 2*—Apply fiberglass roping at the rate of 250 yards per 100 square feet of wall area.

d. *Operation No. 3*—Apply 6 mil coat of polyester coating.

e. *Operation No. 4*—Sand lightly to remove projection of glass fibers.

f. *Operation No. 5*—Apply final 6 mil coat after curing.

POLYESTER WALL SURFACING MATERIAL OR EQUAL (ALTERNATE TO DIVISION 26)

DIVISION 26 (A)

General Notes

1. *Surface Preparation*

All surfaces must be dry and free from dirt, dust, oil and grease.

Brick, plaster, board, trowelled plaster and cement asbestos board should first be primed with Globe No. 18 Latex Primer or equal.

2. *Coating Systems*

Since the beauty and durability of Polyester is enhanced by the application of relatively high mil thickness coatings, it is recommended that the complete Polytile Surfacing systems be applied as follows:

a. Surfaces subjected to water and high moisture conditions such as showers, bathrooms, etc.
   1) Polyester Base Color brushed at approximate rate of 75-100 square feet per gallon.
   2) Polyester Base Color sprayed to 15 mil film thickness. May be applied directly over the wet brushed surfaces. The Brush Coat and Spray Coat should give approximately 30 mils of Base Coat. The first coat is brushed to assure filling the voids in surface. The Brush and Spray Coats combined should give about 50 square feet per gallon.
   3) Apply flecks or dots with one or more Base Colors as desired. Coverage will vary depending on degree of flecking but should consist of about 500 square feet per gallon using two colors as shown in color samples. Flecking may be applied immediately after the Base Coat.
   4) Polyester Glaze Coat sprayed at 6 mil film thickness. The Glaze Coat may be applied at any time after the Base and Fleck Coats have dried hard. This is normally 2½ to 3 hours at 80 degrees F. It is recommended that the Glaze Coat be applied within 24 hours of the application of Base and Fleck Coats.

3. *Mixing and Application*

The Contractor shall meet the Manufacturer's Standards and Specifications for mixing and application of Polytile or equal.

Note: Application of Polyester is not accepted in temperatures below 50 degrees F or above 100 degrees F. Direct sunlight during the curing process is prohibited.

4. *Cleanup*

The Contractor shall follow the Manufacturer's Specifications for cleanup.

1. General

Provide quarry tile work where shown or scheduled on drawings and herein described.

a. Before installing floors, the concrete slab underneath shall be broom cleaned and thoroughly washed and brushed with a cream of neat cement and water, after which the concrete fill is to be applied. All work is to be laid in strict accordance with the basic specifications of the Associated Tile Manufacturers except as herein varied.
b. Concrete setting beds, where required to bring the surface to within 1½" of the top of the finish floor, will be furnished and put in place by the General Contractor.
c. All interior tile base shall be installed by the "Thin Setting Bed Method" using organic adhesives manufactured especially for this type of installation and meeting all the requirements of the U.S. Department of Commerce, Bureau of Standards CS-181-52 specifications.

2. Samples

Typical samples of the kinds and grades of tiles specified and proposed to be used shall be submitted to the Architect of Engineer for approval. Each sample shall be marked with the name and manufacturer and the grade of the tile. Approved samples shall be retained by both the Architect or Engineer and the tiling contractor.

3. Cleaning

Upon completion, all exposed tile and accessory surfaces shall be thoroughly cleaned in a manner not affecting surface finishes.

4. Protection

Before traffic is permitted over finished tile floors, floors shall be covered with building paper. Board walkways shall be laid on floors continuously used as passageways. Floor areas to be trucked on shall have suitable continuous plank runways installed over the building paper. Cracked, uneven, broken or damaged tiles shall be removed and replaced.

5. Location of Quarry Tile Floors and Base

a. Rooms 128, 100 and 106.

COMPOSITION FLOORING AND COUNTER COVERING

DIVISION 28

1. Work Included

a. The work to be done under this heading includes the furnishing of all labor material, equipment and services required for the installation of all composition and resilient floors and counter tops indi‹ ated on the darwings and/or the Room Finish Schedule and specified herein.

2. Materials

a. *Asphalt Tile*—First quality, 1/8" gauge, 9" x 9" size, colors shall be equal to Vicoat Terrazza Series.
b. *Vinyl Counter Tops*—(Classrooms) shall be 0.080 gauge Delux Coat Vinyl or approved equal in colors as selected by the Architect or Engineer.
c. *Vinyl Base*—4" Vinyl Base as manufactured by Vicoat or equal in standard color shall be furnished and installed where specified in Room Finish Schedule.
d. *Metal Edgings*—(Note details where wood edges are used.) Anodized aluminum 3/4" drop below flange, tapered flange, concealed fastenings, similar to Trim edge A-831-S. All necessary vertical and horizontal metal cove members shall be installed as required.
e. *Adhesives*—All adhesives used for installation of tile flooring shall be of waterproof type as recommended by the manufacturer of the tile used.

3. Patterns and Colors

a. Colors of floors and counter tops shall be selected by the Architect or Engineer from C range of colors.
b. Kindergarten room shall have alphabet letters and numbers, complete alphabet a thru z and numbers 1 thru 9 and 0. Color and location within room is to be selected by the Architect or Engineer.
c. Multi-purpose room to have basket ball court using feature strips as shown on drawings.
d. Provide shuffleboard pattern in multi-purpose room, two complete sets (4 triangular designs).

4. Installation

a. All resilient tile shall be installed in areas as scheduled on the Room Finish Schedule in strict accordance with the manufacturer's recommendations.
b. No resilient tile should be installed until interior painting has been completed, except where special coatings are called for.
c. Remove grease and dirt from sub-floors. Inspect floors for holes, cracks, and smoothness; do not proceed with laying until sub-floor is in satisfactory condition. Maintain 70 degree F minimum temperature in rooms for 24 hours before and during time of laying tile and for 48 hours after laying. Stack tile in rooms at above temperature 24 hours before laying. Lay tile in accordance with recommended specifications of manufacturer, and use only experienced workmen. Lay tile with joints tight and in true alignment.
d. After floors have set sufficiently to become seated, wash with neutral cleaner, and thoroughly buff. Upon completion, leave floors and base clean, smooth and free from buckles, cracks, projecting edges, and clean of all mastic.
e. Edges for cabinet tops listed on Plan and detailed shall be applied in maximum widths with adhesive as recommended by the manufacturer and with tight butt joints.
f. This Contractor shall provide metal strips to butt tile to where tile ends adjacent to floor that receives no tiling.

5. *Guarantee*

This Contractor shall guarantee that all work executed under this section of the Specification will be free from defects in material and workmanship, provided any such defect is brought to the attention of the Contractor in writing within one year after completion of the work. Upon such notice, the Contractor shall, at his own expense, make the necessary repairs or replacements of the defective work in question.

1. General

The work done under this heading is subject to the General Conditions, Division 1.

2. Material–All Glass

All glazing for exterior or interior doors and all windows and borrow lights shall be clear glass DSB.
All glazing, unless noted elsewhere, shall be clear glass DSB.
Mirrors are specified elsewhere.

3. Glazing

The Contractor shall set all glass and mirrors in such a way that there shall be an equal bearing the entire width of each pane, except plate glass, which shall have two bearings. Glass that is set with wood or metal stops shall be set without putty but shall have a putty back set to prevent rattling.

4. Putty

Whitting putty shall be used unless otherwise specified. Use putty as it comes from the container without adulteration. All putty work shall be left smooth and free from marks and other defects. The Contractor shall be held responsible for any defective putty work or improper setting. Glazing in steel openings shall be done with special steel sash putty. Glazing clips shall be used for all glazing in steel sashes.

5. Cleaning

The Contractor shall thoroughly clean all glass, both inside and out, before he leaves the job. The window cleaner shall take special care not to deface the putty and painted surfaces and shall exercise extreme caution not to scratch glass surfaces.

1. General

In areas where an acoustical ceiling is designated, in finish schedule, this Contractor shall furnish all material and all labor necessary for installation of the specified ceiling.

2. Material

a. Suspended Grid System
   1) Tile shall be Dura Fire or equal, in 24" x 48" x 1/2" panels. Panel shall have class A fire rating.
   2) Suspension Grid System shall be Acoustical Fire Guard Exposed Grid System or equal.
   3) Fire guard (or equal) lay in panels shall be coated with Mylan facing or equal in the following rooms: 101, 102, 103, 109, 110, 117, 120, 122, and 123.

3. Application

Application shall be as recommended by Dura Ceiling Systems or equal manufacturer.

a. Suspended Grid System shall be suspended with No. 9 wire from joist. The system shall be installed in strict accordance with manufacturer's recommendations and in such a manner as to achieve the specified fire retardant time-design rating. Systems shall be installed to permit border units of the greatest possible sizes.

4. Procedure of Work

The General Contractor shall work with a subcontractor to insure a properly installed ceiling. The General Contractor is responsible for the ceiling after installation until acceptance of building by Owner.

1. Cash Allowance

Include the amount of twenty-three hundred dollars (2,300) for the purchase of finish hardware. This is the net amount available for purchase of finished hardware by the general contractor. It is exclusive of profit, overhead, handling and installation charges, which amount shall be included in the contract price.

2. Selection

a. Selection of finish hardware will be made by the Architect or Engineer. Finish hardware will be available to the general contractor in due time, so as not to interfere with construction progress.
b. Should Architect's or Engineer's selection of finish hardware exceed the amount of the above allowance, owner will pay excess costs. Should Architect or Engineer selection of finish hardware be less than the allowance above, the general contractor's contract will be adjusted accordingly and full credit passed on to the Owner.
c. Architect or Engineer retains option to obtain informal bids on finish hardware and to approve supplier.

3. Items Excluded

Items of hardware not included in the above allowance and to be furnished by others:

a. Window hardware

4. Items Included

The following is a rough list of finish hardware listing various items of hardware to be used at respective locations. The purpose of this list is only to aid the General Contractor in estimating the cost of installation. This list should not in any way be taken as the final complete hardware list.

Exterior Doors

Butts, 1½ pair
Locks
Closers
Hold open devices
Kick plates
Stops

Other Doors

Butts
Locks or latches
Closers as required
Stops

PAINTING AND FINISHING DIVISION 32

1. General Notes

The work under this contract shall include all labor and materials for painting and decorating, including the necessary preliminary preparation of all surfaces designated on the accompanying drawings and hereinafter specified.

2. Materials

a. *Manufacture and Delivery*–All material shall be made by responsible manufacturers approved by the Architect or Engineer as hereinbefore stipulated. They shall be delivered to the work in sealed containers and manufacturer's certificate as to quantities purchased shall be delivered to the Architect or Engineer when requested. No alteration or mixing of materials on premises will be permitted, excepting as specified for priming, sizing or undercoats, and where tinting is required to produce proper color.

3. Colors and Samples

The Contractor for this work shall, at least thirty days in advance of undertaking same, submit to the Architect or Engineer for his approval a complete schedule of the manufacturers of all products required throughout the work, together with a description of the special processes or methods of application recommended by the manufacturer in each case. It is understood, however, that no general approval by the Architect or Engineer of such a schedule shall constitute a waiver of any specific requirements of the specifications. The Architect or Engineer may require, if he deems it advisable, a specific guarantee from the manufacturer regarding the quality and composition of his product.

The Contractor for this work shall, at least thirty days in advance of undertaking same, secure necessary instructions from the Architect or Engineer and prepare for their approval a sample of every color and finish required in the work. Samples shall be adequate in size and made upon materials identical with those to be finished under the contract. Samples shall be resubmitted until approved by the Architect or Engineer.

4. General Requirements

a. *Workmanship*–All workmanship throughout shall be highly skilled such as to develop the fullest possibilities of the material and processes specified.

b. *Protection of Work*–This Contractor shall be fully responsible for the protection of his own work and that of other Contractors, from injury or staining. He shall provide the necessary drop cloths and cover all finish surfaces adjacent to his work. Special care shall be taken to avoid coating or spattering hardware, light fixtures, and finish floors. All damage done by smearing with paint and varnish shall be promptly and completely repaired.

   1) In case of all work done by painter, he will be held responsible for results reasonably to be expected from the materials and methods specified.
   2) Painter shall also be responsible for the surface to which paint is applied, and after application of same, no excuse of improper surfaces shall be accepted. In all work covered by the painter and especially in the case of plaster painting, painter shall go over his work after completion and repair all damaged stops whether due to defective materials or workmanship or defects of the surfaces covered. The necessary additional coats of paint shall be applied to cover all spots or discoloration of every sort without additional charge.

c. *Time of Drying*–Time of drying shall in all cases be ample to secure the best possible results.

5. *Preparation of Surfaces*

a. *Woodwork*–All nail holes shall be carefully filled with colored putty, all knot holes, pitch pockets or sappy portions shall be sealed with shellac. All raised grain shall be sanded smooth. Where water or acid stain is specified, a brush coat of water shall be applied to raise grain, and this shall be sanded before the stain is applied.

b. *Iron Work*–All stain and rust shall be removed and surfaces wiped clean.

c. *Galvanized Iron*–All galvanized iron required to be painted shall be prepared with a solution containing 6 ounces of copper acetate to the gallon of water, applied 24 hours in advance of painting.

6. *Acceptance of Surfaces*

Before commencing painting, the painter must be certain that the work to be covered is in perfect condition to receive the paint; that the surface is clean, dry, smooth and at the proper temperature. Should the painter find surfaces and conditions impossible of acceptance, he shall at once report such conditions to the Architect or Engineer and cease operation on the portion of work affected. The application of paint shall be held to be an acceptance of the surface and working conditions, and the painter shall be held responsible for the results reasonably to be expected from the materials and process specified.

7. *Priming and Backpainting*

a. *Shop and Mill Priming*–All coats herein specified shall be in addition to shop and mill priming elsewhere specified.

b. *Contract Surfaces*–Contract surfaces will not be required to be painted except as follows: All contract surfaces of wood or metal window frames, where same abutt against masonry walls, shall be painted one good coat oil paint just before being brought into contact. This coat is in addition to mill priming.

8. *Exterior Work*

a. Exterior woodwork shall be painted two coats, in addition to prime coat, of an approved exterior paint. Caulking shall be painted.

b. All exterior metal work is to be painted two coats exterior paint in addition to a Galvanized metal prime coat. Sheet metal work on roof shall be painted. Copper flashings to be left unpainted.

c. All exterior door frames and structural steel should receive in addition to Prime coat, two coats lead and oil paint.

9. *Interior Work*

a. Metal work (all exposed metal members except aluminum): Aluminum does not require painting. Unfinished electrical cover plates in ceiling shall be painted. Exposed ductwork, grilles, registers, louvers, etc., shall be painted.

b. Door Frames are to be painted, in addition to their prime coat:
   1) One coat half enamel, half undercoat
   2) One coat enamel (Stain finish)
      (Last coat enamel in toilet rooms is to be gloss)
   3) Lightly sand between coats

c. All exposed structural steel lintels and joists etc., are to be painted, in addition to their prime coat:
   1) One coat lead and oil paint
   2) Finish coat lead and oil paint

d. *Masonry Work*—All interior lightweight block masonry walls scheduled to be painted shall have:
    1) Two coats of Poly Paint Vinyl Seal or Tester and Treats Vinyl Masonry
    2) Paint shall be applied by brush or spray and as recommended by manufacturer for application to lightweight masonry units
e. All wood doors, trim and cabinet work:
    1) First coat—1/2 filler and 1/2 stain consisting of 2/3 Driftwoor and 1/3 Pecan P&L, applied and wiped
    2) Second coat—Shellac, water clear, sanded and puttied
    3) Third coat—Gloss varnish
    4) Fourth coat—Dull Varnish
f. White Pine shelving portions of cabinets are to have:
    1) One coat shellac, water clear
    2) One coat varnish, flat
g. All factory-primed door grilles, louvers, registers, convectors and fin pipe covers shall be painted as follows:
    1) Two coats of semigloss topcoat corresponding to adjacent area colors
h. The top and bottom portion of all interior surfaces of all wood cabinets and drawers shall be sealed.
    Note: All unit ventilators have a baked enamel finish and require no painting.
i. Exposed pipe covering shall be painted one coat of Pond "Sealer Coater" plus one coat of topcoat corresponding to the adjacent areas.
j. Viewable portions of interior of ductwork shall be painted flat black.
k. Underside of expose decking in Multi-purpose room shall receive:
    1) Two coats of base and one of vinyl masonry

10. Door Numbering

Stencil at the top of one side of all interior room doors, three digit numbers (101, 102, etc.). Numbers shall be 1-½" high with approximately 3/16" stroke width. The numbering system shall be as directed by the Architect or Engineer.

11. Door Lettering

Stencil on one side of the door, 1" below the numbers, the following 2" high lettering using 1/4" stroke width:

Rooms	*Lettering on Door*
Boys Toilet and Boys Locker room	1 each Boys
Girls Toilet and Girls locker room	1 each Girls
Office	Office
Supply-Storage	Supply, Storage
Kindergarten	Kindergarten
Multi-purpose room	Multi-purpose

12. Materials and Application

a. Paint for above numbers and letters shall be white gloss enamel.
b. Style of numbers and letters shall be "Ribbon" style, and may be either installed by decal method, stencil, or "sign painter" applied.

1. Chalkboards and Tackboards–Work Included

The work under this specification shall include the furnishing of all labor, materials, equipment and services necessary for the furnishing and installation of all chalkboards and tackboards as shown on drawings and as specified.

a. *Materials–Chalkboards and Tackboards*–Chalkboards shall be similar and equal to Clear Products & Equipment Inc., No. 100, 1/2" thick Clearcity Green color or Weber and Stalk Co's Sterling 5/32" thick "litesite" Green color. Tackboards shall be natural cork mounted on hardboard. Two Clear or equal No. 953A, 3' x 5' bulletin boards shall be furnished and installed in locations designated by the Architect or Engineer.
b. *Installation of Chalkboards and Tackboards*–All chalkboards and tackboards shall be installed in exact accordance with the manufacturer's printed instructions. After installation the surface must be cleaned to remove paint, glue, etc., and then broken in as recommended by the manufacturer in his printed instructions. Tackboards shall be installed so that the finished face is flush with the chalkboards face plane. Trim shall be prefinished and masonry walls behind boards shall be painted in accordance with manufacturer's instructions.
c. *Chalkboard Guarantee*–The General Contractor shall furnish to the Architect or Engineer duplicate copies of a certification that installation to all of the chalkboards is in accordance with this specification and the manufacturer's recommendations. The supplier shall furnish duplicate copies of the manufacturer's printed guarantee certifying to the fact that all chalkboards were perfect at the time of shipment and guaranteeing the chalkboard installation and the writing surface against defects of workmanship and material for the life of the building.

2. Flagpole

a. *General*–Furnish and install tapered aluminum flagpole as shown on Plans.
b. *Material*
    1) Tapered aluminum flagpole 40' 0" from grade to top and 4' 0" in ground
    2) Concrete-pier as shown on Plans
    3) Aluminum flashing collar

3. Alternate G-5: Basketball Backboard and Backboard Mounting Devices

a. *General*–All basketball backstops shown on plans or covered by the Specifications shall be as manufactured by Gibbs Equipment Corporation, Anderson, Indiana, and distributed by Josline Brick and Supply of Omaha, or approved equal.
b. One Wall Mounted Stationary Model 120 constructed on 1½" I.D. pipe and malleable iron fittings. Cross bracing 1/4 x 1¼ flat mild steel. Guy rod 3/8" diameter steel with turn buckles. Douglas Fir wall pads finished in two coats clear varnish. Aluminum finish on all fittings. Furnished with complete wall anchors.
c. *Backstop*–Shall be 14 FM and MG goal and 60 thread net–steel single sheet 12 gauge steel. Backboard has smooth 1½" wide rolled edges and welded reinforcing channels; the finish is egg shell non-glare enamel.

4. Alternate G-2

If this Alternate is accepted, the Contractor is to furnish and install Folding Tables of the recess type shown on the drawings. Tables and frame shall be equal to Wall-up Model 16 in the wall type as manufactured by Wall Table, Inc., 10 Melrose Place, Los Angeles, Calif. Bench length will be 7' 6"; table height 2' 5". The Supplier shall provide a brochure giving construction and specifications of the table and benches with a recessed frame for the Architect's or Engineer's approval.

5. Alternate G-3

If this Alternate is accepted, the Contractor is to furnish and install a steel canopy as shown on the drawings. The canopy shall be equal to Alon pre-engineering walkway covers by Alon Manufacturing Company, P.O. Box 87, Houston 8, Texas. Structural design shall be in accordance with the standards and specifications of the American Iron and Steel Institute, 1961 Edition. Assembly shall be straight and plumb. Footings are to be provided by the General Contractor.

DIVISION 34

This particular division is to be used for any specifications of material and labor required to install a product that is not mentioned in any other part of the specifications, and must be purchased and installed by the contractor for usage by the owner. These items would be the mechanical things such as a prefabricated walk-in cooler, or a drive-in depository window for a bank. Each of these items is manufactured by specialty manufacturers and usually installed by the manufacturer in cooperation with the contractor's craftsmen.

1. General Notes

a. The bidder shall carefully examine the drawings and specifications and visit the site of construction. The bidder shall fully inform himself to all the existing conditions and limitations, and shall include in his bid a sum to cover the cost of all items necessary to complete the project.
b. The bidder shall be familiar with the city, county, state and national codes, and to which code takes precedence, and where the drawings and specifications do not cover such items, the bidder shall be responsible for them.
c. It is pointed out that the general conditions and special conditions of the specifications are considered a part of these specifications, and this division.
d. Should bidder find discrepancies in, or omissions from the drawings and specifications, or should he be in doubt as to their meaning, he should at once notify the Architect or Engineer, who will send written instructions to all bidders. Neither the Owner nor the Architect nor Engineer will be responsible for any oral instructions.
e. Provide all items, articles, equipment, material, operations, and tools for the erection of the mechanical systems shown on drawings or specified herein, including labor supervision, and incidentals required and necessary to complete the systems for successful operation. Install systems in complete accordance with standards of local, state or national codes governing such work. Place complete systems in a safe and satisfactory condition; adjust all automatic control devices in prescribed operation.
f. All equipment installed in strict accordance with manufacturer's recommendations and these specifications unless otherwise approved in writing. Vertical and horizontal dimensions of all fixtures, units and equipment suitable for installation in spaces provided therefore. Ample clearance provided for maintenance and repair. Instruct Owner in operation and maintenance; furnish one set to the Owner of repair parts list and one set instructions on operation and maintenance all equipment installed. All material new and unused.

2. Fees and Permits

a. Pay all fees, permits, taxes for inspection, utility charges on connection, etc., in connection with the contract. Upon completion of work, furnish owner with certificate of final inspection from inspection bureau having jurisdiction.

3. Drawings

a. Drawings forming a part of these specifications are numbered and described as follows:
Sheet P-1 Plumbing
P-2 Plumbing

4. Data and Measurements

a. Date herein and on drawings is exact as could be obtained. Absolute accuracy not guaranteed. Obtain exact locations, measurements, levels, etc., at site and satisfactorily adapt work to actual building conditions; installing system generally as shown on plans.

5. Delivery and Storage of Material

a. Provide for storage and safe delivery of materials and arrange with other Contractor for introduction into building of equipment too large to pass through finished openings. Deliver materials at such stages of work as will expedite work as a whole and mark and store in such a way as to be easily checked and inspected.

6. *Materials and Equipment*

a. When make and type definitely specified, bids must be based on that make and type. When specified on two or more types, equally acceptable, one of make and type mentioned, but that make and type throughout. When specified a certain make or type, "or approved alternate," Make and type specifically required unless written approval of "alternate material" is obtained from Architect or Engineer, prior to delivery on job.

7. *Approval Data*

a. Approval granted on shop drawings and manufacturer's data sheets rendered as a service only and not considered guarantee of quantities, measurements or building conditions; nor construed as relieving the contractor of basic responsibilities under contract.
b. For items differing from those specified herein and requiring approval, the Contractor shall submit complete data within 30 days after contract award. Failure to submit required approval data within the 30 days after the contract award shall be interpreted to mean that the contractor will furnish equipment as specified. Requests for approval of substitute materials after the 30 days period will not be considered unless it can be shown that, due to circumstances beyond the Contractor's control, the specified material is not available and cannot be obtained in time to prevent unnecessary delay of construction.
c. For items of approval, the Contractor shall submit six (6) sets of complete schedules of material and equipment proposed for installation. Include catalogs, diagrams, drawings, data sheets and all descriptive literature and data required to enable the Architect or Engineer to determine compliance with construction and contract requirements. Schedules and all data submitted at one time; no consideration will be given to partial or incomplete schedules.
d. Include with data, complete analysis describing all differences between substitution and item specified; describe changes required for piping, ducts, wiring space, structure, and all portions of project affected by substitution.
e. The Contractor will be limited to one approval on a singular piece of equipment. Consideration will not be given where the Contractor submits approval data on one piece of equipment, by more than one manufacturer.

8. *Installation of Work*

a. Examine drawings and specifications for other work, and if any discrepancies occur between Plans for this work and plans for work of others, report such discrepancies to Architect or Engineer and obtain written instructions for changes in work of others. Any changes in work, made necessary through neglect or failure of Contractor to report such discrepancies, made by and at expense of Contractor. Confer and cooperate with others on work and arrange this work in proper relation with theirs. Contractor will be held solely responsible for proper size and location of anchors, inserts, sleeves, chases, recesses, openings, etc., required for proper installation of work. Arrange with proper contractors for building in anchors, etc.; do all cutting and patching made necessary by failures or neglect to make such arrangements with others. Any cutting or patching subject to directions of Architect or Engineer and not started until approval obtained. Damage due to cutting shall be repaired by the Contractor. Notify General Contractor of exact sizes and location of all openings required for recessed units and equipment items.

9. *Electrical*

a. Furnish all motors, automatic or manual controls, starters, relays, protective and signalling devices required for operation of equipment specified herein; furnish complete wiring diagrams and instructions to Electrical Contractor for installation. Wiring diagrams and instructions shall be in hands of Electrical Contractor within two weeks after contract award. Safety disconnect switches, where required, shall be provided by the Electrical Contractor.
b. Manual Control shall be Dallen Bulletin 600 or 609 starters of proper horsepower and voltage rating. Motors of any rating may be directly controlled by automatic control devices such as thermostats, aquastats, float switches or pressure

switches, where such controls have adequate horsepower and voltage, rating and proper protection for overloads of motors and are incorporated within the control device. Otherwise the control unit shall be utilized for the operation of the starting and holding coil of a Bulletin 709 starter.

10. Temporary Water Service and Toilet Facilities

a. Temporary water service and toilet facilities are the responsibility of the General Contractor, and all facilities shall be at the expense of the General Contractor. However, the General Contractor may request the service of the Mechanical Contractor for installation of the Temporary facilities for use during the period of construction, and also for the removal of the equipment when the construction is completed. As soon as construction permits and upon direction of the Architect or Engineer, install either permanent or temporary toilet facilities connected to the plumbing systems, within the building, for use during construction.

11. Trenching and Backfilling

a. Trenches excavated true to line and grade, banks properly sheeted, shored, braced where required. Provide accurately contoured bottom for uniform support on undisturbed soil for at least 1/3 pipe circumference for each section of pipe along its entire length, except bell holes excavated progressively with pipe laying. Care should be taken in excavating that foundations or footings are not injured in any way. Control grading and stacking to prevent surface water flowing into trenches.
b. Any water accumulating shall be removed continuously. Backfilling shall not be done until pipe joints are thoroughly set. After testing lines and inspection, backfill trenches with excavated material free of debris, large clods or stones, 4" layers, uncompacted thickness, moistened and thoroughly tamped, until pipe has covered at least one foot. Remainder excavated material, in 6" layers, moistened and tamped to density undisturbed earth. Flooding trench shall not be permitted. Should subsequent settlement occur, trench shall be opened to depth required, refilled, and compacted. Pavement or walk cuts shall be restored to original condition.

12. Piping Installation

a. *General*–Cut to measured fit at building, installed parallel to walls and ceiling, properly clear all openings; provide required clearance as for operation of doors, windows, access panels, valves, etc.; excessive cutting not permitted. Make changes in direction with fittings, except bends permitted in soft temper tubing. Interior pipe shall be kept free of cuttings, dirt scale and loose material of any nature. Open ends shall be capped or plugged when work is not in progress. Install to permit free expansion without causing damage to joints, pipe hangers, units to which connected, structure, or undue noise. Exposed piping shall be run closely as possible to finish surfaces, piping consealed in wall chases or spaces or furring provided wherever possible. Piping shall be tight and tested before enclosed. Unions shall be made where required for disconnection to facilitate quick repair without disconnection of long lengths or equipment. Valves, trap cleanout, or other part normally required for operation or maintenance not installed in inaccessible place.
b. Drainage and Vent Piping
    1) General: Minimum grade horizontal drainage piping shall be 1/4" per foot for 3" and less 1/8" per foot 4" to 8", 1/16" per foot for larger. Change size drain lines with reducing fittings or recessed reducers; change direction with 45 degree wyes, half wyes, long sweep quarter bents. 1/6, 1/8 or 1/16 bends, except sanitary tees may be used on vertical stacks and short quarter bends or elbows may be used in drainage lines where flow from horisontal to vertical.
    2) Underground: Pipes laid hubs facing up grade. Not less than two lengths of pipe in position, joints finished, earth fill tamped along side pipe, ahead of each joint before poured, except at closure.
    3) Above Ground: All main vertical soil and waste lines shall be extended full size to and above roof as vents. Where practicable, two or more vent pipes connected together, extended as one pipe through roof, 6 pound sheet lead

extending on roof 8" from pipe in all directions, carried up, over, into top of pipe 1½" or with Deploy No. 400 elastic flashing, Dekimical Company, in accordance with manufacturer's recommendations. Vents extended above roof 12", increased in size 12" below under side of roof as required. Minimum size through roof 4". Vent pipes in roof spaces and under floors shall run close as possible to construction, horizontal runs pitched down to stacks without forming traps. Connection of end or circuit vent pipe to vent serving other fixtures, at least four feet above flood level fixture served. Branch waste connections to fixture same size fixture outlet unless otherwise shown.

13. Joints and Connections

a. *General*–All joints shall be water and gas tight. Caulking threaded joints or holes will not be permitted. Paint, varnish, putty, etc., will not be permitted on joint until joint is tested tight. Crosses used on vent pipes, or as specified. Joints in steel pipe shall be made with clean cut threads, graphite and oil compound applied to make thread only, pipe reamed full size after cutting. Joints in hard copper tubing hard soldered surfaces properly cleaned and bright, butt ends reamed, fluxed as recommended by maker of fittings. Flared joints shall be made with tubing expanded without splits, using proper flaring tool. Concealed joints in copper water piping, sweat, hard soldered, or brazed. Type of solder for fittings shall be in accordance with data in the National Bureau of Standards Publication, "Building Material and Structures Reports BMS58 and BMS8," as tabulated by the Copper and Brass Research Association.

b. For prevention of electrolytic corrosion at connections between pipe of dissimilar metals, such as steel to copper, provide dielectric pipe unions of flange unions similar to Lanco, 3204 Sackett Avenue, Cleveland 9, Ohio.

14. Drainage Piping

a. Pipe or fittings with double hubs on same run or double tee branches will not be permitted on soil or waste piping. Drilling, tapping or welding soil, drain or vent pipes or use of saddle hubs or bands will not be permitted. Dead ends shall be avoided except where necessary to extend cleanouts to accessible elevations. Slip joints in drainage piping shall be used only in waste pipe between trap seal and fixture. Metal contact unions used in drainage piping only in trap seals and on inlet side trap.

b. Tile pipe shall have two rounds well tamped rope oakum; fill balance with cement mortar rammed in position, wiped flush with hub. Interior joints shall be wiped clean with swab large enough to fill pipe, drawn past each joint as made. Mortar, 2 parts fine sand to one of cement and after set 1/2" thickness coal tar pitch troweled over entire joint. Joints cast iron to tile pipe encased with heavy mass 1/2 portland cement mortar, minimum 4" thickness around pipe, 18" long. Clay pipe joint compound CPI-2 used in lieu of mortar for joints, option of Contractor. If used, compound shall be heated and poured in strict accordance with manufacturer's recommendations, including application of coal tar film-forming primer whenever wet, oily or cold conditions prevail.

c. Cast iron pipe, picked oakum and lead, 12 ounces lead each inch diameter in each joint made, poured full at one pouring and caulked solid. Joints lead to cast iron pipe, with extra heavy brass ferrules, extra long, wiped to lead pipe, caulked into hub cast iron coupling screwed to steel pipe to and from spigot end.

15. Cleanouts, Test Tees

a. Cleanouts at end of each horizontal drainage run or change in direction, at each branch connection, at intervals not over 40 feet in horizontal runs. Each cleanout provided with brass ferrule and brass screw plug, chromium plated cast brass handhole where necessary in finished room. For under flue or drains, cleanouts branch brought to finished floor level with long sweep 1/4 or two 1/8 bends, brass screw plug with counter sunk spud set flush and level 4" in larger lines, 6".

b. All cleanouts shall be located and arranged and be easily accessible for rodding. Test tees with brass screw plug installed 30" above floor at foot of all drainage.

16. Hanger, Supports

a. Support piping firmly, prevent sagging, pocketing, swaying or displacement, making due allowance for structural requirements. Suspended pipe shall be held by vertical adjustable split ring pipe hangers. Metal pipe covering protectors where necessary to prevent deformation of insulation. Chain, wiring or perforated hangers will not be acceptable. Trapeze hangers may be used in lieu of separate hangers, spacing per smallest pipe. Chromium-plated pipe shall be supported by chromium-plated cast brass supports. Hook-plates for pipe supported from side walls. Support vertical waste vent or drainage pipes at base of stack. Install hangers in line adjacent to hangers and in accordance with the following schedule, except hub and spigot pipe 5' or less in length supported 5' centers, suppott close to hub. Vertical piping supported each floor.

Pipe Size, inches	*Minimum Spacing Hangers, feet*
1/2"	6
3/4 and 1	8
1/4 and 2½	10
3 and 4	14
6 and 8	18

17. Floor, Wall, Ceiling Plates

a. Uncovered piping passing through floors, finished walls or finished ceilings fitted with plates or escutcheons large enough to completely close openings around pipe, securely held in place, plated finish. Caulk water tight around pipe in unfinished room.

18. Sleeves, Inserts

a. Provide sleeves secured in place, sufficient size to accommodate pipe passing through masonry or concrete. Sleeves in structural supports, galvanized steel pipe. Elsewhere, either galvanized sheet metal or especially designed fiberboard sides with metal caps and washers, Caneral Company, Vallejo, California. Sleeves through floors flush with finished floor and space between pipe and sleeve caulked water tight. Contractor to set sleeves and inserts in drilling finished work.

19. Supports, Fastenings

a. Fixtures, equipment secured to wood with round head brass screws, to masonry with brass bolts or machine screws, lead sleeves type and anchorage units, or 1/4" brass expansion bolts. Exposed portions of fastenings shall be chromium plated. Support fixtures hung from flock walls with concealed wall hangers attached to iron fish-plated at rear of wall, long through bolts, nuts, and washers.

20. Traps

a. Each fixture, all equipment requiring drain connection, equipped with trap same size fixture outlet, placed as near fixture as possible. Water closed traps integral with ware.

21. Sterilizing

a. Cold, hot water lines, storage tank and equipment, sterilized with sufficient chlorine to provide dosage not less than 30 P.P.M., contact period 16 hours, all valves opened and closed three times during sterilizing period. Following contact period, water thoroughly flushed from system until residual chlorine content not more than 0.20 P.P.M.

22. Cleanup, Adjusting

a. All parts of work left clean; equipment, fixtures, valves, pipe and fittings cleaned of grease and metal cuttings. Any discoloration or other damage to portions of building, its finish or furnishings, due to Contractor's expense. All automatic control devices adjusted for proper operation. All surplus materials and rubbish shall be removed as it accumulates. All equipment shall be left in safe and proper operating condition.

23. Tests

a. Drainage and venting system shall be filled with water to level of highest stack and proved tight in presence of Architect or Engineer. If necessary to cover pipe before roughing is ready for inspection, tests shall be made with water at 15 psi. Cold, hot water pipe tested 100 psi, proved tight.

24. Insulation

a. All work shall be performed by an experienced insulation contractor. All piping shall be clean, dry and tested prior to insulating. Sectional covering secured in place with outward clinching staples of sufficient number to properly secure the covering, 4" centers minimum. All open ends of pipe insulation shall be neatly finished off with cement. Contractor shall be required to remove any insulation showing evidence of deterioration and replace with new insulation as directed. Exposed piping in rooms, including equipment rooms, shall have 6 ounce canvas jacket pasted neatly over vapor barrier.
b. *Domestic Cold Water*–Insulate with 1/2" thick Dacon Snap on, or equal, glass fiber sectional pipe insulation with factory applied vapor barrier jacket of aluminum foil adhered to 55 pound minimum white kraft vapor paper, moisture vapor transmission not over 0.03 perm, and joints strips and overlap seams adhered with vapor barrier mastic.
c. Domestic hot water lines, insulated with 3/4" thick Dacon Snap on, or equal, glass fiber sectional pipe insulation with factory applied "Universal" jacket or standard canvas jacket neatly pasted in place.

25. Guarantee

a. This Contractor shall guarantee all materials, workmanship, and the successful operation of all apparatus installed, and to furnish service free of charge on all portions of the systems, for a period of one year from date of acceptance of construction, and shall repair any defect, or replace apparatus within a reasonable time after notice thereof at his own expense; provided such defect is in the opinion of the Architect or Engineer, a fault of the equipment and not due to the misuse of the the equipment.

26. Pipe and Fittings

a. Drainage pipe installed 5' or more outside building lines, No. 1 salt glazed tile vitrified, sewer pipe. Drainage and vent pipe from 5' outside and all such pipe inside of building of sand service weight cast iron, hub and spigot. Asphaltum coated conforming to Commercial Standard CS188-59, except fixture wastes above ground and 2" and less in diameter and branch vents unless otherwise shown, may be galvanized steel pipe with cast iron coated drainage type fittings.
b. Water pipe above ground, hard drawn copper tubing, type L, with wrought copper fittings, sweat joints. Water pipe not laid in trench with sewer or drain pipe.

27. Water Valve Stops

a. 125 psi water working pressure, gate valves of non-rising stem type with wedge disconnect threaded joints, packing and

disconnects suitable for service intended, brass or bronze body. All valves 3" and less shall be
Valves 3/4: and smaller may be globe valves; all others gate valves. Check valves of 45 degree swing type or angle plated loose key stops on supplies to all fixtures.

28. *Fixture Connections*

a. Floor supported water closets secured to heavy cast brass floor flange with brass bolts and chromium-plated face nuts; floor flange anchored to floor with brass lag screws or brass bolts set in lead sleeves. Joint between bowl and piping shall be sealed gas and water tight by graphite asbestos gasket.
b. Exposed traps and supply pipes for all fixtures and equipment shall be connected to rough pipe at wall unless otherwise specified.
c. Water connections to individual fixtures not less than the following:

Water closets	1¼"
Urinals	3/4"
Service sink	1/2" hot and cold
Classroom sinks	1/2" hot and cold
Lavatories	1/2" hot and cold
Drinking fountains	1/2" cold
Wall hydrants	3/4" cold
Showers	1/2 hot and cold
Kitchen equipment	As shown on plans

29. *Fixtures and Equipment*

a. All fixtures designed to prevent backflow of polluted water or waste into water supply system. Plating may be of nickel or chromium on polished surface. All fittings on fixtures shall be same surface finish. Flush valve handles shall be chromium plated. Chromium-plated stops and supplies with plated escutcheons for all fixtures. Mounting heights of lavatories, drinking fountains and sinks shall be as directed. Immediately after fixtures are set, cover with manilla paper glued on; also assure trim is adequately protected. Provide guards and boxing as necessary to protect fixtures from normal operations of other trades. Before installing fixtures, all water piping shall be thoroughly flushed out to remove all dirt, oil, chips or other foreign matter. Upon completion of the work, this Contractor shall remove protective covering from fixtures and thoroughly clean and polish fixtures and trim. All fixtures shall be Valor or equal.
   1) Lavatory (L) Valor No. 1623 Am 19" x 17" vitreous china with cast in soap dish, wall hangers, supply pipes with stops to wall 1¼" P trap with cleanout. Combination supply fittings with pop-up drain. Furnish complete with supply. Mount 27" above floor.
   2) Water Closet (WC-1) Valor 4280ET, siphon jet bowl with elongated rim 1½" top spud inlet with Cloan Royal 110FYV exposed flush valve with vacuum breaker, 1" screw driver stop, seat bumper Thurch No. 9500 Black open front east with check hinge.
   3) Urinal (U) Valor K 4920TB2, 18" wide vitreous china floor mounted with integral flush spreader, strainer for 2" drain connection, 2" P trap. Cloan Royal No. 185H exposed flush valve with vacuum breaker and screw driver stop. Provide 2" Urinal seam covers for continuous battery installation.
   4) Sink (S) Valor Model K-5991A 20" x 24", 8" deep with 4" ledge; single compartment, acid resistant enameled cast iron, strainer with non-removable grid, 1½ P trap with cleanout No. K7830 Utilla mixer faucet, stainless steel rim and lugs.
   5) Service Sink (SS) Valor K6652-A, K8895, pail hook and wall trap K6673, 3" enameled inside with strainer. 22" x 20" with 10" back and 28" from floor to rim.
   6) Shower Valor K7255 build-in-shower, 1/2" Valor mixer, renewable seats, K7384 ball joint shower head with volume regulator.

30. Floor Drains (FD)

a. In floors on earth, with trap, integral or adjacent cleanout, have cast iron body with brass grate, threaded brass cleanout plug, spigot outlet. Floor drains in finished rooms to have chrome-plated brass strainers and in unfinished rooms to have plain brass strainers. Indirect waste drains Royal Series 30037Es. All brass floor drains with funnel strainers. Provide drains with special offset rims Royal Type AF where required for floor coverings.

31. Wall Hydrant

a. Royal 1411 Series or equal, cast brass 3/4" non-freeze wall hydrant, 3/4" hose connection, polished brass face, key handle and brass wall sleeve fitted with brass locknut.

32. Drinking Fountain (DF)

a. Valor K5390 A recessed unit with non-squirting bubbler head, self closing control valve adjustable for continuous flow. Automatic volume regulator screw driver regulating stop. 3/8" I.P. connection strainer, brass trap with cleanout excention to wall.

33. Roof Drains

a. Roof drains, 4" Fremont, installed as indicated on plans and as recommended by manufacturer.

34. Water Heaters

a. Furnish and install Fairfax circulation heaters as shown. Furnish with Type AR thermostat and connect heating elements through contractor. Contractor shall be Dallon 100 amp, three pole 208 volts. Each unit to be 30 kw at 208 volts three phase. Units are 48½" high, 8-5/8" diameter with 2" outlet and inlet connections.

b. Hot water storage tank, 42" diameter, 96" long, 576 gallon capacity, working pressure of 125 lb/in./sq. ASME-NB inspected and labeled U.L. approved. Tank heads 285 -c flange quality steel tank sheet 385 -c flange quality steel. Galvanized inside and out after fabrication and test. Tank to be insulated after installation with 1½" joints and staggered, secured with galvanized annealed steel wire over insulation 2" hexagonal mesh wire stretched and ends tied together. Finish with two coats cement. First coat to dry before second coat is applied. Cover with 3 ounce canvas cemented with arabol lagging adhesive 60-89-05. Paint with aluminum paint. Tanks to have hubs as necessary to complete the plumbing installation. Contractor to install with No. 140 x Watts temperature and pressure relief valve, shutoff valve at tank inlet, and 3/4" Drain valve.

35. Kitchen Equipment

a. All kitchen equipment will be furnished by others, including traps, valves, strainers, or other components necessary for installation. This Contractor shall rough in for all equipment and make final connections to the following items:

Disposals (2)
Pot sink
Peeler
Dishwasher
Booster heater

Information for all equipment (kitchen) will be furnished by the General Contractor.

36. Circulating Pumps

a. Provide 3 Spinner Circulating Pumps all with flanged-in line fittings, two will be 2" flanges and one 1" flange, each will be 1/6 h.p. single phase motor. Two pumps will be mounted on the supply of the Circulating heaters, and shall be controlled by aquastats mounted on the return tank feed; such aquastats shall be adjusted for 140 degrees. One pump shall be mounted on the return of the domestic circulating line and controlled by an aquastat mounted on the supply line, which will be set at 140 degrees F. Contractor shall furnish aquastats.

1. *Scope*

The work under this heading shall include furnishing and installing all materials and equipment specified and shown on the drawings for the Heating System.

2. *Gas Piping*

a. *Material*–All gas piping shall be standard weight, mild steel, black pipe. Pipe 2" and larger shall be welded. All screwed fittings, except cocks and valves shall be standard weight, beaded, malleable cast iron. Valves shall be brass, ground key, plug cocks. All underground piping shall be coated and wrapped.

b. *Installation*–All piping shall be reamed to full size after cutting and all screwed joints shall be made up with red lead applied to the male threads only. All pipe shall be run true to line without pockets and with even pitch to a suitable point where an approved drain cock shall be provided.

c. No unions shall be used in concealed piping. All outlets not connected to equipment or appliances shall be closed with malleable caps.

d. Upon completion, the entire gas piping installation shall be tested under and proven tight at an air pressure of 50 pounds gauge per square inch.

3. *Steam and Condensate Piping*

a. *Material*

1) All steam supply piping shall be standard weight, Schedule 40, black mild steel pipe, Onida Steel, or equal.
2) All condensate return piping shall be standard weight, genuine wrought iron, onida Company, or equal.
3) All pumped condensate piping shall be standard weight, genuine wrought iron, Onida Company, or equal.
4) Threaded fittings shall be gray cast iron, standard weight, free of sand holes and imperfections with clean American Standard taper pipe threads, complying with Federal Specification SS-P-501, 125 pounds.
5) Steel and wrought iron welding fittings shall conform to ASA-B16-9-1951 Steel Butt Weld Fittings.
6) Tee connections shall be made with standard tee fittings. Where wrought iron piping is welded, wrought iron fittings shall be used.

b. *Installation*

1) *Steam Mains* shall be run as shown on the drawings and shall be evenly pitched not less than 1" in 50 feet in the direction of steam flow. The mains shall be of sizes indicated and shall run as shown on the drawings. Use eccentric reducers where reduction of main sizes occurs in the direction of steam flow in horizontal mains. The end of steam main shall be dripped into return line through trap.
2) *All Return Mains* shall be of sizes shown and shall be evenly pitched not less than 1" in 50 feet in direction of flow.
3) *Allowance for Expansion* shall be made in the installation of all piping so that the usual variation in temperature will not cause undue stress at any point. Pipes shall be securely anchored where necessary to properly distribute expansion stresses.

   All branch mains shall be supported in such a way as to permit expansion and contraction and to relieve runouts of all weight.
4) *Joints*–All pipe 2½" or larger shall be welded as set forth in the Standard Manual of Pipe Welding of the Heating, Piping, and Air Conditioning Contractors Association. Piping 2" and smaller shall be screwed.
5) Where pipe is welded, flanges shall be installed at equipment connections and as required to disassemble piping in equipment rooms for maintenance work.
6) Valves, traps and specialties for the steam piping shall be as specified hereinafter.

4. *Valves*

The Contractor shall furnish and install all valves of the sizes, pattern and make indicated on the drawings and described in these specifications. All valves 2" and smaller shall have brass bodies. All steams shall be properly packed. Valves 2" and smaller shall be threaded. Valves larger than 2" shall be brass trimmed, iron body type, unless otherwise specified. Valves 2½" and larger shall be flanged.

a. *All gate valves 2" and smaller* shall be wedge disc pattern, with non-rising stem for not less than 125 psi steam working pressure, or equal.
b. *All gate valves 2½" and larger* shall be wedge disc non-rising stem for not less than 125 psi steam working pressure and shall be Palmer, or equal.
c. *All globe valves* shall be for not less than 125 pounds steam pressure, similar and equal to Palmer for 2" and smaller and No. 884 for 2½" and larger.
d. *All angle valves* shall be for not less than 125 pounds steam pressure, similar and equal to Palmer.
e. *All check valves* shall be horizontal swing type for not less than 125 pounds steam pressure, similar and equal to Palmer-1200 up to 2", and No. 1790 for 2½" and larger.

5. *Traps*

a. *Thermostatic Traps*–Thermostatic traps shall be provided for all finned tube radiation, similar and equal to Gates, rough brass body, sized for the capacities of each unit as indicated on the drawings.
b. *Low Pressure Float and Thermostatic Traps*–Low Pressure Float and Thermostatic Traps shall be provided at ends of low pressure mains and at all air handling coil units, similar and equal to Gates. Thermo Traps shall have minimum capacities shown on the drawings at 2 psi pressure difference.

6. *Strainers and Unions*

a. Strainers shall be "y" pattern, copper strainer, iron body, similar and equal to Copperweld.
b. Unions shall be provided adjacent to each screwed type valve and shall be on the outlet side of the valve. Unions in steel shall be heavy duty malleable, screwed type with brass to iron ground joints. Unions in copper tube shall be copper with solder connections and copper to copper ground joints.

7. *Pressure Gauges*

a. Provide pressure gauges, U.S. Gauge, or approved equal, with white dial and black scale, size 6" dial. These gauges shall be located for easy reading. Each gauge shall be equipped with an integral or separate siphon and shall be connected by means of a brass pipe and fittings containing a shutoff cock.
b. Steam system pressure gauges shall be as follows:

Steam gauge	0-30 lb graduation
Condensate pump	
Discharge gauge	0-30 lb graduation

8. *Boilers*

Furnish and install steel, fire-tube type steam heating boilers as shown on the drawings. Boilers shall be arranged to burn fuel hereinafter specified. Boiler shall be ASME Code designed and stamped for a 15 psig steam working pressure. Boiler shall be gastight with gasketed doors and forced draft burner.

a. Boiler shall be installed complete with the following items:
   1) Refractory lined combustion chamber
   2) Steam pressure gauge
   3) ASME Rated pressure relief valve
   4) Gauge glass with trimmings and dry-cocks
   5) Insulated metal jacket with factory finish
   6) 3/4" hose bibb drain

b. *Burner*—Boiler shall be arranged to burn natural gas.
   1) Burner shall be designed to burn natural gas efficiently; shall be completely automatic and shall be a complete unit with power type blower and air switch. Burner shall be installed in strict accordance with manufacturer's recommendations and the requirements of the Local Gas Company.
   2) Burner controls shall consist of a steam pressure control to cycle the burner operation, an air switch wired through the burner control circuit to hold the gas valve closed, unless the switch has been actuated by the burner fan. A high limit control to stop the burner and close the gas valve if the steam pressure reaches a predetermined high limit. Ignition shall be gas electric, intermittent. Flame failure protection shall be electronic type, with photoelectric cell and flame rod supervision of main and pilot flames.
   3) All control wiring, line and low voltage required for burner operation shall be furnished and installed by this Contractor.

9. Boiler Foundations

Each boiler shall be set on a concrete pad, isolated from the floor by an asphaltic expansion joint. The top edges of the pad shall be chambered 3/4". The foundation shall be installed as recommended by the boiler manufacturer. The pad shall have No. 4 reinforcing bars on 12" centers both ways. The pad shall be 8" thick.

10. Fin Tube Radiation

Furnish and install fin tube radiation of the type, lengths, dimensions and capacity shown on the drawings and herein specified. All components of the fin tube installation, except the heating elements, shall be given a phosphatized or bonderized treatment to prevent rust and shall be finished with a baked, gray prime coat. Radiation shall be installed in strict accordance with the manufacturer's recommendations. The fin tube radiation shall be manufacturer indicated or equal. Installation shall be complete with covers, end pieces, hangers, brackets, access doors, corners, trim strips and pilaster covers as required for a neat installation.

a. Fins shall be nonferrous and bonded to copper tubes by mechanical expansion of tube.
b. Hangers shall be of sliding bracket type. Covers shall be of 16 gauge steel and shall be firmly supported at top and bottom on exnclosure brackets at maximum intervals of 6 feet. A compressible sealer strip shall be installed between the enclosure and the wall. Access doors shall be provided where required for access to valves and vents.
c. Covers shall extend from wall to wall except as otherwise indicated. Elements shall be evenly spaced below cover.
d. All lengths of fin tube element assemblies exceeding 12 feet shall have expansion joints as specified and of same size as pipe.
e. Fin tube ratings shall be approved under the I-B-R Code for fin tube type of radiation.
f. Delete fin tube radiation in classrooms for alternate H-1.

11. Connectors

Convectors shall be of the manufacturer shown or equal. Convector cabinets shall be of die-formed 16 gauge furniture steel, and of the type shown on drawings.

a. Cabinets shall have prime coat and shall be rustproofed.
b. Convectors shall be rated in accordance with Commercial Standard US140-147 by National Bureau of Standards, U.S. Department of Commerce.
c. Convector elements shall be nonferrous, mechanically bonded, rigidly constructed throughout, pitched for drainage and properly vented.
d. Convectors shall be installed as shown on the drawings with capacities.

12. Propeller Unit Heaters

Furnish and install propeller type unit heaters of the manufacturer shown, or equal, as indicated on the drawings.

a. Heaters shall be enclosed in furniture steel cabinets properly braced and stiffened for maximum rigidity with factory finished heat resistant enameled exterior. Heating coil shall be nonferrous, free to expand and contract, minimum tube interior diameter 3/8 inch, hydrostatically tested to 200 psi for maximum working pressure of 125 psi. Fans shall be quiet in operation, direct driven with motor shielded from the coil, resilient mounted, ball bearing, with overload protection.
b. Heaters shall be rated in accordance with Industrial Unit Heater Association Code.
c. Capacities shall be as indicated on drawings.
d. Horizontal discharge heaters shall have defle ors for both horizontal and vertical directional air control.
e. Control shall be as indicated in "Temperature Control."

13. Cabinet Unit Heaters

Furnish and install, where shown, cabinet unit heaters of manufacture shown or equal.

a. Units shall be floor mounted as shown, with capacities, characteristics and unit size as shown in the schedule. Units shall operate quietly.
b. Install valves and accessories as detailed.
c. Heaters shall be enclosed in 16 gauge furniture steel cabinets properly braced and stiffened for maximum rigidity, with factory finished, heat resistant enameled exterior. Heating coil shall be nonferrous, free to expand and contract, hydrostatically tested to 200 psi for maximum working pressure of 125 psi. Motor shall be shielded from the coil, resilient mounted, ball bearing, with overload protection.
d. Control shall be as indicated in "Temperature Control."

14. Expansion Joints

Expansion joints shall be Expanso flexible metal expansion compensators, designed for use in the heating system, and shall be suitable for installation in copper or steel pipe heating lines as required. Joints shall be installed where shown on the drawing.

15. Gas Vents

a. Boiler and water heater vents and breechings shall be Ventroid double wall gas vent, or equal.
b. Installation shall conform to all recommendations of the manufacturer and the National Board of Fire Underwriters. Vents shall be flashed and counterflashed at the roof line and shall have approved type weatherproof caps.

16. Heating and Ventilating Unit

The unit shall be furnished and installed as indicated on the drawings.

a. Unit shall be complete with casing, fans, motor, belt drive, belt guard, coil, filters and dampers for face and bypass, fresh air and return air.
b. *Casings*—Each air handling unit shall be housed in a casing constructed of steel sheets, not less than 0.0478 inch in thickness (18 gauge). The cabinet shall be adequately reinforced and stiffened with steel angles or other structural members and shall be provided with all necessary interior panels, supports for equipment, access openings and dampers. Casing openings connected to ducts shall be equipped with removable angles and bolts for attaching canvas or other flexible connections. All interior surfaces of the casing shall be rendered rust-resistant. The cabinet shall be insulated on the inside with not less than ½" of moisture-resistant, moldproof, vermin-proof insulation. Removable panels in the casing shall provide easy access to all parts for lubrication and servicing when the unit is installed.
c. *Heating Coil,* unless otherwise specified, shall be extended-surface nonferrous alloy construction, with the fins securely bonded to the tubes. The coils shall have been tested hydrostatically and proved tight under a gauge pressure of 150 psi. The coils shall be properly pitched to permit complete drainage and shall be encased in a galvanized metal frame. The coils shall have capacity not less than that indicated on the drawings. Coil shall be steam distributing non-freeze type.
d. *Circulating Fans* in unit assemblies shall be of the centrifugal, multiblade type. Each fan unit shall have an air capacity not less than that indicated on the drawings. Each fan unit shall be installed complete with electric motor and drive equipment. The fans shall be rated and constructed in accordance with the AMCA Standard Test Codes. Fans shall be statically and dynamically balanced at all speeds. The fan shall have two externally mounted self aligning, grease lubricated, ball bearings. The fan shafts shall be made of steel, and shall be provided with key seats and keys for the impeller hubs and fan pulleys, or with other equally positive fastening. V-belt drives shall be designed for at least 50 percent overload capacity. Each fan motor shall be equipped with adjustable base or rails for belt tightening. Motor mountings shall be on a resilient base constructed of steel shapes and connected through rubber-in-shear.
e. *Motors* shall be squirrel cage, general purpose 40 degrees C, polyphase NEMA frame, electrical characteristics as indicated in the schedule. Starter will be furnished and installed by the Electrical Contractor. Wiring to the motor lugs by the Electrical Contractor.
f. *Air Filters*—All filters shall be 2" thick throwaway type Filter or equal. Size and capacity to be coordinated with the units.
g. *Mixing Box*—Combination filter and mixing box shall be installed as indicated on the drawings. Each opening shall be provided with a damper and control rods—see "Temperature Control." Mixing box shall be factory assembled unit.
h. *Control of Unit* shall be as specified in "Temperature Control."
i. Unit shall be provided with the necessary piping, valves and accessories, as indicated on the drawings and as specified.

17. Unit Ventilators: (Alternate M-2 Only)

Furnish and install complete factory assembled unit ventilators. The units shall be of the size, capacity and manufacturer shown on the drawings.

a. *Casings* shall be constructed of 16 gauge furniture steel, properly braced and formed for maximum rigidity. Casings shall be bonderized and given a factory prime coat.
b. *Ratings and Capacities* shall be as indicated on the drawings. Units shall be arranged for the control cycle specified in Section "Temperature Control." Unit ventilators shall be tested and rated in accordance with the ASHRAE Standard Code for Testing and Rating Unit Ventilators.
c. *Fans* shall be forward curved, double-inlet, centrifugal type. Fans shall be provided in sufficient number to insure quiet

operation with a maximum noise rating of 50 decibels. Any unit found in the opinion of the Architect or Engineer to be excessively noisy shall be removed and replaced at the expense of this Contractor.

d. *Coils* shall be nonferrous, with silver brazed or mechanically bonded connections. Capacity shall be in accordance with schedule shown on plans.

e. *Motors* shall be capacitor type, totally enclosed, mounted on a resilient base, wound for 115 volt, single phase, 60 cycle service, with thermal overload protection.

f. *Filters* shall be 1" thick, renewable figerglas type with metal frames and fiberglas pads.

g. *Electric Wiring* for units shall be by Electrical Contractor to junction box inside unit. Interior wiring to be by this Contractor. Starting switch will be furnished by unit manufacturer.

h. *Protective Covering*–All unit ventilators and accessory items shall be shipped with a protective covering so applied that access panels can be removed without disturbing this cover. This Contractor shall be responsible for repairing to the satisfaction of the Architect or Engineer any damage incurred during installation.

i. *Fresh Air Intake Louvers in Masonry Walls* shall be manufacturer's standard design complete with insect screen and aluminum stamped ornamental intake grille. Furnish to General Contractor for installation.

j. Provide one unit ventilator under each classroom window.

18. Condensate Return Pump

Furnish and install one duplex condensate return pump with capacity and characteristics shown on the drawings. The pump shall be of the manufacturer shown on the drawings, or equal. The pump shall be driven by open dripproof ball bearing motors with the electrical characteristics indicated. The pump shall be capable of the indicated operation while handling condensate to 200 degrees F. The equipment shall include one cast iron receiver tank, two pumping units, inlet, vent, drain and pump discharge connections, strainer and a mechanical alternator to allow alternate pump operation and simultaneous pump operation if required. A gate valve and check valve shall be installed in each pump, and a gate valve on the inlet to the receiver. Provide a pressure gauge in the common discharge line. The receiver vent shall be galvanized steel pipe installed full size through the roof and properly flashed. The vent may be connected to a common vent if so shown.

19. Water Feeder

Furnish and install, on each boiler, a combination water feeder and low water cutoff switch of a capacity equal to the evaporation rate of the boiler, Miller & Jones No-182, or equal.

20. Boiler Water Treatment System

Furnish and install for the boiler feed system a pot-type feeder having a minimum capacity of 2 quarts and constructed for an operating pressure of 150 psi. The unit shall be complete with shutoff and isolating valves, flow control valve, and drain valve. The feeder shall be installed on the discharge side of the condensate pump. A shop-fabricated feeder construction from wrought iron may be used if approved. All piping shall be wrought iron with cast iron fittings.

21. Cleaning of Boilers and Piping

After the hydrostatic tests have been made and prior to the operating tests, the boiler shall be thoroughly and effectively cleaned of foreign materials. The boilers shall be filled with solution consisting of 1 pound of Sal Soda (Sodium Carbonate) per 30 gallons of water and operated for a period of 24 hours, then drained-flushed and refilled with fresh water.

1. Scope

The work under this heading shall include furnishing and installing all materials hereinafter specified or shown on the drawings, for the ventilating system.

2. Roof Exhausters–Centrifugal Type

Roof exhausters–The centrifugal type shall be installed as shown on the drawings and shall be of the type, wheel size and capacity indicated on the drawings. Exhaust fans shall be of the manufacture shown on the drawings, or equal. Fans shall be of the centrifugal type, belt driven or direct driven as shown on the drawings. V-belt drives shall be designed for 50% overload and shall have adjustable pitch motor sheaves. Motors shall be mounted on adjustable bases.

Housings shall be fully weatherproofed, and all metal parts exposed to the air stream shall be given a sprayed-on insulating undercoating. Outlets shall be provided with removable bird screens. Units shall be mounted on approved vibration isolating bases.

Fan bearings shall be of the ball bearing type and provided with adequate and accessible means of lubrication. Each unit shall have a disconnect located in the motor compartment. Fans shall be quiet in operation. Fan housings shall be easily removable for access to all parts.

3. Bathroom Exhaust Fans

Furnish and install residential type bathroom exhaust fans where indicated on the drawings. Fans shall be of the centrifugal type as shown on the drawings. Fans shall be quiet in operation. Fans shall have gravity type dampers. Fan shall be of the capacity shown on the drawings and of the manufacture shown, or equal.

4. Roof Curbs

Roof curbs shall be of the prefabricated metal type and designed for mounting roof exhausters and air intakes furnished. Curbs shall be of the sound attenuating type with insulated lining. Curbs shall be designed for flashing to the roof in an approved manner. Curbs shall have adequate means for mounting dampers where dampers are specified or shown on the drawings.

HEATING AND/OR VENTILATING DIVISION 35c

1. General

The work required under the heating and ventilating contract shall include all material, labor, and equipment to completely install with proper trades the system ready for use as outlined in these specifications and as indicated on the Plans. All work, materials, and manner of placing material to be in strict accordance with latest requirements of the manufacturer, National Board of Fire Underwriters, and with local and state laws and ordinances relating to this work. All material shall be new, unused, of high quality in every respect, and shall conform to standards of ASHRAE, and established standards for the particular type of equipment or material.

2. Drawings

Drawings forming a part of these specifications are numbered and described as follows:

Sheet No. H-1

3. Data and Measurements

Data herein and on the drawings is as exact as could be obtained. Absolute accuracy is not guaranteed. Obtain exact locations, measurements, levels, etc., at the site and satisfactorily adapt this work to actual building conditions, installing systems generally as shown or indicated.

4. Delivery and Storage of Materials

Provide for delivery and safe storage of materials. Deliver materials at such stages of work as will expedite work as a whole and mark and store in such a way as to easily checked and inspected.

5. Materials and Equipment

a. When make and type definitely specified, bids must be based on that make and type.
b. Approval granted on shop drawings and manufacturer's data sheets rendered as service only and not considered guarantee of quantities, measurements or building conditions; nor construed as relieving Contractor of basic responsibilities under contract.

6. Installation of Work

Examine drawings and specifications for other work, and if any discrepancies occur between plans for this work and plans for work of others, report such discrepancies to the Architect or Engineer and obtain written instructions for changes necessary to accommodate this work to work of others. Any changes in work, made necessary through neglect or failure of the contractor to report such discrepancies, will be made by and at the expense of the Contractor. Confer and ccoperate with others on wo and arrange this work in proper relation with theirs. The Contractor is held solely responsible for proper size and location of anchors, inserts, sleeves, chases, recesses, openings, etc., required for proper installation of work. Arrange with the proper contrac tor for building-in anchors, etc. and for leaving required chases, recesses, openings, etc.; do all cutting and patching made necess by failure or neglect to make such arrangements with others. Any cutting or patching is subject to the directions of the Architec or Engineer and not started until approval is obtained. Damage due to cutting will be repaired by this contractor. Notify the general contractor of the exact sizes and locations of all openings required for recessed units and equipment items.

7. Electrical

The Electrical Contractor shall provide all material and labor to wire all equipment and controls. The Heating and Ventilating Contractor shall furnish to the Electrical Contractor all motors, automatic or manual controls, starters, relays, and protective signaling devices required for operation of the equipment specified, and furnish complete wiring diagrams and instructions within two weeks after contract award. Safety disconnect switches, where required, shall be provided by the Electrical Contractor.

8. Duct Work

a. Duct work must be permanent, rigid, non-buckling, and nonrattling with all flat sides cross broken. Joints in duct work shall be airtight. Galvanized iron sheets of the following U.S. Standard gauges shall be used in the construction of all duct work.

Greatest Dimension	*U.S. Gauge*
0" to 12"	26
13" to 30"	24
31" to 60"	22
61" to 90"	20
Over 90"	18
Diameter, inches	*Gauge*
6" to 19"	26
20" to 29"	24
30" to 39"	22
40" to 49"	20
Over 49"	18

b. All changes in the direction of airflow and in cross sectional area shall be accomplished with well designed, properly installed streamlined fittings. Supply ducts must be securely supported by metal hangers, straps, lugs or brackets. No nails shall be driven through duct walls and no unnecessary holes shall be cut in them.

c. The size of the ducts shall be as indicated on the plans and shall be net inside dimensions of the sheet metal duct.

d. The general location of the ducts and flues shall be as indicated on the plans; exact locations shall be determined at the building and the Contractor shall furnish and install such additional bends and offsets as may be required to bring the duct work into proper relation with other equipment and features of the building.

e. Where thermal duct insulation is called for on the Plans, the insulating material shall be vapor barrier faced Thermo Duct Insulation of the thickness specified as manufactured by Thermo Insul Company or approved equal. The insulation shall be held in place with spot daubings of a quicktacking, rubber-based adhesive on approximately 6" centers. All end and longitudinal joints shall be butted firmly and sealed by taping with a 2" wide vapor barrier pressure sensitive tape. Tape ends are to lap at least 4". Plain vapor barrier tapes applied with adhesive may be substituted.

f. The return air round duct shall be a prefabricated glass fiber duct system of the size shown on the plans as manufactured by Thermo Duct Manufacturing Company or approved equal. The duct system shall be installed in accordance with the manufacturer's recommendations as outlined in their manual for installing Thermo Duct.

9. Registers

All supply registers and return grilles shall be of the size and type shown on the Plans.

10. Gas Piping

A gas service line shall be provided from the L.P. gas storage tanks to the heating units in the Mechanical Room. Gas piping shall be installed in accordance with the requirements of the State of Nebrasks, the local L.P. gas company and the

National Board of Fire Underwriters rules and regulations. Gas piping shall be black steel pipe with malleable iron fittings. A lubricated plug gas stop valve and gas pressure regulator shall be installed in the line at the building entrance. Burners shall be provided with manual shutoff, a pressure regulating valve, 100% automatic gas shutoff valve, and safety pilot assembly. Gas piping shall be adequate to provide enough fuel to each burner to develop the nameplate output of the appliance served during normal operation of all L.P. gas burning appliances in the building.

11. L.P. Gas Storage Tanks

Provide and install two 1000 gallon steel storage tanks for L.P. gas meeting the requirements of the State of Nebraska, local L.P. gas company and the National Board of Fire Underwriters. Each tank shall be equipped with a high pressure valve and each unit using gas from these tanks shall be equipped with a low pressure valve. The tanks shall be interconnected so that gas from each tank will be available to the system automatically. The tanks shall be equipped with a liquid level gauging device of approved design.

12. Guarantee

a. The Contractor, by accepting this specification and signing of the Contract, acknowledges familiarity with all requirements and guarantees that every part including equipment, fittings, sheet metal, pipe, work etc., used to make up the system herein provided, will be of the highest quality in its field and will be erected by skilled and experienced workmen in a thorough and substantial manner.
b. This Contractor shall guarantee all materials, workmanship, and the successful operation of all equipment and apparatus installed by him for a period of one year from the date of final acceptance of the whole work, and shall guarantee to repair or replace at his own expense any part of the apparatus installed by him for a period of one year from the date of final acceptance of the whole work, and shall guarantee to repair or replace at his own expense any part of the apparatus which may show defect during the time, provided such defect is, in the opinion of the Architect or Engineer, due to imperfect material or workmanship and not carelessness of improper use.

13. Testing an. Operating

a. After the plant has been completely installed the Contractor shall operate the entire system for five (5) days and will thoroughly clean and blow out the system to remove all dirt, sand, scale, grease, lint, etc., before final acceptance. Before turning the work over to the Owner, the Contractor shall have all parts tested and in good working order and adjusted to operate as intended by the specifications.
b. All equipment bearings and motor couplings shall be lined up true and even so that there is no thumping, knocking, or noise present when the equipment is operating at full rated capacity. When aligned, all equipment shall be securely anchored in place.
c. The Contractor shall adjust all supply, return, and exhaust openings to provide the amount of air called for on the Plans. Velocity readings shall be made with a velometer, anemometer, or other approved velocity meter in an approved manner and the readings recorded as directed. Where the amounts of air are not shown on the drawings, the velocity through the openings shall be equalized. After preliminary adjustments are made, the system shall be checked with thermometers and supply ducts to each room adjusted so that all temperatures and equalized
d. All thermostats shall be adjusted and set to operate as intended.
e. All air filters shall be clean.
f. The Contractor shall instruct the Owner on operation and maintenance of the system during the 5 day operating period.
g. The Contractor shall furnish all labor required for the test; the Owner will furnish electrical current and fuel required for the test. The Contractor shall thoroughly lubricate the system during the test and leave the system completely lubricated upon completion of the test and before the system is accepted.

14. *Plumbing*

The Plumbing Contractor shall provide and install a floor drain for removing condensate from the Direct Expansion Coils located in the furnace plenums.

15. *Equipment and Controls*

a. The heating and ventilating units shall be of the model and size as shown on the Plans and hereinafter schedule. Manufacturers' name and equipment numbers given in the following specifications are for the purpose of establishing *minimum standards,* and equipment by another manufacturer, *completely meeting such standards,* will be acceptable.

b. Schedule of Furnaces

1) *Basement System*

1 a) Furnace: Model 360-Hytemp Furnace 360,000 BTUH Input and 288,000 BTUH Output. Complete with filters and 2¾ hp Fan Motors and fan capable of supplying 5080 CFM at 0.4" Static Pressure.

b) Controls: Provide a Dyna low voltage heating/cooling thermostat at the location shown on the plans. The furnace fans shall operate when exhaust fans in the Basement are operating. The furnace fan motors are to be wired so that both furnace fans will run during heating or ventilating operations.

c) Fresh Air Intake: Provide Cast Wall Louvers equipped for power operation of the size and at the locations shown on the Plans. The Louver Damper shall be equipped with a bug screen and controlled to open when exhaust fans in the Basement are operating and close when exhaust fans in the Basement are not operating.

d) Provide Venta flues as shown on the Plans.

2) *First Floor System*

a) Furnaces: 2 Model, 288 Hytemp Furnaces each unit having 360,000 BTUH input and 288,000 BTUH Output. Complete with filters and each furnace shall have 2¾ hp Fan Motors and fans capable of supplying 5080 CFM at 0.4" Static Pressure.

b) Controls: Provide Dyna low voltage heating/cooling thermostats at the locations shown on the Plans. The furnace fan motors shall be wired so that both fans in each furnace will run during heating or cooling period. The furnace fans shall operate when exhaust fans on the first floor are operating.

c) Fresh Air Intake: Provide Cast Wall Louvers equipped for power operation of the size and at the locations shown on the Plans. The louver damper shall be equipped with a bug screen and controlled to open when exhaust fans on the first floor are operating and close when exhaust fans on the first floor are not operating.

d) Provide Vent flues as shown on the Plans.

c. Schedule of Exhaust Fans:

Model	Make	Capacity CFM	Location	Remarks
176	Spinner	1760	Lounge "B"-113	Provide Size 12 Intake Guard, Size 12 Automatic Wall Louver & Size 12 Motorized Wall Louver
299	Spinner	2990	Lounge "A"-112 & Meeting Rm. 111	Provide Size 18 Intake Guard, Size 18 Automatic Wall Louver and Size 18 Motorized Wall Louver
225	Turner	225	Men #1-105	Provide 838-AL Wall Cap 7" R Duct-Model 860 Switch
300	Turner	300	Womens #1-107	Provide 838-AL Wall Cap 3¼" x 10" Duct-Model 860 switch
299X	Spinner	2990	Mens Locker Rm #1-124 & Mens Locker Room #2-128	Provide Size 18 Intake Guard, Size 18 Automatic Wall Louver, and Size 18 Motorized Wall Louver
225CF	Turner	225	Mens R.R.-127	Provide 885-AL Wall Cap 7" R Duct-Model 860 Switch
250	Turner	250 ea.	Card Room-118	Provide roof ventilators
180	Turner	180	Womens R.R.-121	Provide 838-AL Wall Cap 3¼" x 10" Duct-Model 860 switch
176	Spinner	1760	Womens Locker Rm-120 & Womens Dress.Rm-122	Provide Size 12 Intake Guard, Size 12 Automatic Wall Louver and Size 12 Motorized Wall Louver

16. Combustion and Ventilation Air

Provide an opening in the top and bottom of the door to the Mechanical Room 116 for combustion and ventilating air. Provide 1 square inch free area for each 1,000 BTU per hour of input. Locate the grille above the opening of the draft hood in the door or wall. Provide, also, a combustion and ventilation air inlet grille with 1 square inch of free area for each 1,000 BTU per hour of burner input. Locate the opening in the wall or door below the combustion air inlet to the furnace and at least 3½ feet center to center of the grilles from the ventilating air outlet grille.

17. Air Conditioning

Furnish 10 ton compressor-condensor unit, and "A" frame coil as manufactured by Hy-temp. Compressor-condensor to be mounted on roof as shown. "A" frame coil to be mounted in furnace Plenum. Extend tubing from unit on roof to coil. Install in line a dryer of adequate capacity. Fill with freon 22, the number of pounds recommended by the manufacturer. Adjust suction and liquid line valves according to instruction with unit.

Provide electrical contractor with necessary diagrams for proper connection of all controls.

Complete installation to be supervised by manufacturers representative and compressor-condensor guaranteed for a period of five years.

1. General

a. The Bidder shall carefully examine the drawings and specifications and visit the site of construction. The Bidder shall fully inform himself to all the existing conditions and limitations, and shall include in his bid a sum to cover the cost of all items necessary to complete the project.
b. The Bidder shall be familiar with the city, county and state and national codes, and to which code takes precedence, and where the drawings and specifications do not cover such items, the Bidder shall be responsible for such.
c. It is pointed out that the general conditions and special conditions of the specifications are considered a part of these specifications, and this division.
d. Should Bidder find discrepancies in, or omissions from the drawings and specifications, or should he be in doubt as to their meaning, he should at once notify the Architect or Engineer, who will send written instructions to all bidders. Neither the Owner, Architect or Engineer will be responsible for any oral instructions.
e. Provide all items, articles, equipment, material, operations, and tools for erections or mechanical systems shown on drawings or specified herein, including labor, supervision, and incidentals required and necessary to complete systems for successful operation. Install systems in complete accordance with standards of local or state codes governing such work. Place complete system in safe and satisfactory condition; adjust all automatic control devices in prescribed operation.
f. All equipment installed in strict accordance with manufacturers' recommendations and these specifications unless otherwise approved in writing. Vertical and horizontal dimensions of all fixtures, units, and equipment suitable for installation in spaces provided therefor. Ample clearance provided for maintenance and repair. Instruct Owner in operation and maintenance; furnish Owner one set repair parts list and one set instructions on operation and maintenance of all equipment installed. All material new and unused.

2. Fees

a. Pay all fees, permits taxes for inspection, utility connection charges, etc., in connection with the contract. Upon completion of work, furnish Owner with certificate of final inspection from inspection bureau having jurisdiction.

3. Drawings

a. Drawings forming a part of these specifications are numbered and described as follows:

Sheet E-1	Electrical
Sheet EHV-1	Heating and Ventilating
Sheet ESS-1	Signal and Alarm Systems

4. Data and Measurements

a. Data herein and on drawings is exact as could be obtained. Absolute accuracy not guaranteed. Obtain exact locations, measurements, levels, etc., at site and satisfactorily adapt work to actual building conditions; installing system generally as shown or indicated.

5. Delivery and Storage of Materials

a. Provide for delivery and safe storage of materials and arrange with other contractors for introduction into building of equipment too large to pass through finished openings. Deliver materials at such stages of work as will expedite work as whole and mark and store in such a way as to be easily checked and inspected.

6. *Materials and Equipment*

a. When make and type definitely specified, bids must be based on that make and type. When specified one of two or more equally acceptable make or types, one of make and type mentioned but that make or type throughout. When specified a certain make or type, "or approved alternate," make and type specifically mentioned unless written approval of "alternate material" is obtained from Architect or Engineer prior to delivery on job.

7. *Approval Data*

a. Approval granted on shop drawings and manufacturer's data sheets rendered as a service only and not considered guarantee of quantities, measurements or building conditions; not construed as relieving contractor of basic responsibilities under contract.
b. For items differing from those specified herein and requiring approval, the Contractor shall submit complete data within 30 days after notice of contract award. Failure to submit required approval data within the specified 30 day period shall be interpreted to mean that the Contractor will furnish equipment as specified. Requests for approval of substitute materials after the 30 day period will not be considered unless it can be shown that, due to circumstances beyond the contractor's control, the specified material is not available and cannot be obtained in time to prevent unnecessary delay of construction.
c. For items of approval, the Contractor shall submit three sets of complete schedules of material and equipment proposed for installation; include catalogs, diagrams, drawings, data sheets and all descriptive literature and data required to enable Architect or Engineer to determine compliance with construction and contract requirements. Schedules and all data submitted at one time; no consideration will be given to partial or incomplete schedule.
d. Include with data, complete analysis describing all differences between substitution and item specified; describe changes required in piping, ducts, wiring, space, structures, and all portions of project affected by substitution.
e. The Contractor shall be limited to one approval on a singular piece of equipment. Consideration will not be given where the Contractor submits approval data on one piece of equipment, by more than one manufacturer.

8. *Installation of Work*

a. Examine drawings and specifications for other work and if any discrepancies occur between plans for this work and plans for work of others, report such discrepancies to Architect or Engineer and obtain written instructions for changes necessary to accommodate this work to work of others. Any changes in work, made necessary through neglect or failure of Contractor to report such discrepancies, made by and at expense of Contractor. Confer and cooperate with others on work and arrange this work in proper relation with theirs. Contractor will be held solely responsible for proper size and location of anchors, inserts, sleeves, chases, recesses, openings, etc., required for proper installation of work. Arrange with proper contractor for building-in anchors, etc., for leaving required chases, recesses, openings, etc.; do all cutting and patching made necessary by failure or neglect to make such arrangements with others. Any cutting or patching subject to directions of Architect or Engineer and not started until approval obtained. Damage due to cutting shall be repaired by the Contractor. Notify General Contractor of exact sizes and locations of all openings required for recessed units and equipment items.

9. *Tests*

a. Entire installation shall be free from short circuits and improper grounds. Test to be made in the presence of the Architect or Engineer. Each panel to be tested with mains disconnected by the Architect or Engineer. Each panel to be with mains disconnected from the feeders, branches connected and switches closed, lamps removed or omitted from sockets, and all wall switches closed. Feeders tested with the feeders disconnected from the branch circuit panels. Each individual power equipment is to be connected for proper operation. In no case shall the insulation resistance be less than that required by the National Electric Code.

10. Rubbish

a. Remove from building, premises and surrounding streets, alleys, etc., all rubbish and debris resulting from operations as it accumulates and leave all material and equipment and spaces occupied by them absolutely clean and ready for use.

11. Construction Power

a. Provide temporary electric service for use during construction complete with metering connections approved by the local power company. Provide by temporary or permanent connections, grounding type outlets, 120/240 volts, 60 cycles, single phase, so located throughout the building that no more than eighty feet of cable will be required to reach any portion of the building, and with the ground pole properly connected to water service pipe or ground rod. Temporary power supply shall be inconnected to water service pipe or ground rod. Temporary power supply shall be installed in such manner as not to endanger life or property. Connections to outlets in the temporary power distribution system shall be made with a minimum of No. 12 conductor. Temporary lighting for construction purposes and payment for all power will be by the General Contractor; temporary protection shall be 20 amp at source of power.

12. Grounding

a. Identified neutral in the interior wiring system and all metallic conduits, supports, cabinets, and equipment permanently grounded to the water system as near as practicable to the point of entrance with approved copper ground clamp. Area of contact at point where wires connect to be sufficient to provide a current-carrying capacity equal to that of the wire. No wires smaller than No. 8 shall be used for grounding. Covering of identified neutral to be white finished, or finished in another, light color. Equipment grounds green finished. Connections protected by conduit fittings. Type and regulation of the National Board of Fire Underwriters and the National Electric Code.

13. Wire and Wiring

a. Type-Wire for general light and power wiring, rubber or the thermoplastic insulated, rate 600 volts, equal to that manufactured by the Electro Company, No. 6 and larger, stranded Conductors soft drawn, annealed copper with conductivity not less than 98%. Insulation "Code" Type TW for branch lighting and power circuits; type RH for feeders. Wire installed in conduit in floors on earth, and underground, "moisture proof" type RHW or RH-RW, unless otherwise noted. Neutral or grounded wire in two or multiple wire branch circuits white finished or otherwise distinctively colored. Multiple circuits coded in accordance with the National Electric Code. Deliver wire on job in original coils bearing manufacturer's name and trademark and Underwriter's label.

b. *Size*–As required and in no case smaller than indicated by diagrams. No wiring smaller than No. 12 accepted for line voltage circuits. Neutral same size as outside wires. On Homeruns where distance to first outlet exceeds fifty feet, minimum size No. 10. For low voltage control switching, and signal systems, the manufacturer of the system shall recommend the wire size to be installed.

c. *Splices, Taps and Terminals*–No splices or tapes permitted except at outlet, junction, pull and panel board boxes. Wires No. 6 and larger connected to panels and apparatus with solderless lugs or connectors. Taps on or splices in wires No. 6 and larger made with solderless connectors. All lugs and connectors are sufficiently large to enclose all strands of conductor and securely fastened. Splices or connections of No. 8 and smaller may be soldered or connected with Scotchlock electrical spring connectors. Where spring connectors are used, insulate with Tenslock insulators, soldered joints or be insulated with rubber and friction or plastic tapes. Insulation should be equal to that of conductor. Prior to making connections, wires shall be cleaned.

d. *Installation*–All wiring shall be installed in metallic raceway unless otherwise permitted by note on plans or elsewhere in these specifications. Feeders, runs in equipment room runs in concrete slabs on earth, and runs in exterior masonry walls shall be installed in rigid conduit. Wiring elsewhere shall be installed in electrical metallic tubing. Wires not pulled until entire conduit system is installed and the building, except for interior finish is substantially completed. No grease, we soap, or other substance harmful to insulation applied to wire. Conductors not pulled until conduit is moisture free. Allow sufficient slack for connections of fixtures, switches, and other devices without additional splices.

14. Raceways

a. Rigid conduit shall be of mild steel, standard weight, sherardized or hot dip galvanized inside and outside, and with smooth enamel coating on interior surfaces. Electrical metallic tubing of mild steel, electro-galvanized, and with smooth enamel coating on interior surfaces. Elbows and fittings shall be of the same finish as the raceway, delivered in ten foot lengths, with maker's name and trademark and Underwriter's label on each length.

b. Underground service from transformer to panel "P" shall be located as shown on the plot plan. The underground portion of the conduits enclosing the service conductors shall be rigid conduits of non-metallic substances buried not less than 24" deep. Non-metallic conduits shall be equal to PVC.

c. Non-metallic conduits shall be adapted to galvanized rigid conduits at each end, and the rigid conduits connected together with a No. 6 bare copper bond conductor brazed to conduits. Adapter at the building end shall be installed five feet from the point where the service enters the building, and conduit extended to service entrance equipment indicated on riser diagrams. Adapter at the supply end shall be installed within five feet of the base of the power company terminal pole or to a height designated by the power company, where it shall be terminated in a weathertight terminal fitting, as furnished by the Power company; conduit from pole to transformer shall be 4" P.V.C. or transite buried 24" deep, and incased in 4" concrete sheath.

d. Service conductors from Pole to transformer slab furnished and installed by Power Company, to be code type RHW, sized as shown on plans.

e. Install as one complete system, with all joints in pipe and connections to boxes electrically and mechanically perfect. Cutting done with hacksaws, and ends out square and reamed. Joints made with standard fittings and watertight. Joints in EMT made with watertight compression type fittings. Indenter type fittings for EMT not acceptable. Changes in direction made with trace elbows, conduits, or pipe bend of radius not less than that of the trade elbow of the same size. All bends made smoothly without crushing pipe or injuring protective coating. Not more than four quarter bends, or equivalent, installed in a line between any two outlet boxes, cabinets or pull boxes. Pull boxes installed whenever necessary to comply with foregoing provision. Raceway to enter boxes squarely, and to be firmly thereto. Exposed raceway to be supported with pipe strips or other approved means, equally spaced not more than eight feet apart. Raceway concealed in concrete or masonry shall be erected ahead of construction and wired in place. Exposed raceway run in a neat and slightly manner, square with walls and ceilings; adjacent runs parallel, at same level, and with concentric bends. Keep raceway systems free of dirt and water, and swab clean before wires are pulled.

f. Raceway shall not be exposed in finished room except as specified or specifically approved by the Architect or Engineer.

15. Outlet Boxes

a. Provide at each current consuming or switching outlet. Boxes set plumb and level, and secured firmly in place with face of box or box cover flush with finished wall or columns. Boxes secured to conduit with galvanized locknuts and bushings, and to tubing with watertight compression type fittings with insulating bushing in throat or fittings. Conduit or tubing offset as required to enter boxes used for pulling. Fixture outlet boxes shall have approved fixture studs fastened to boxes. Outlet boxes for systems other than power distribution system shall be as required by manufacturer of equipment installed.

16. Position of Outlets

a. *Note:* Prior to rough-in:
   1) Obtain approval from Architect or Engineer of outlet locations shown or changes required therein.

2) Consult with mechanical contractor to determine where boxes are to set to avoid interference with mechanical equipment on walls or ceilings.

b. The right is reserved to change outlet locations shown, up to the time of rough in, without change in contract price. Outlets centered with respect to paneling, furring, trim, etc. Where several outlets occur in one room, they are to be symmetrically arranged. Outlets set plumb or horizontal, secured firmly in place. The face of the box extending to finished surface of wall, ceiling or floor as the case may be, without projecting or being recessed. Unless otherwise indicated or required, install outlets at heights noted below, dimensions being in inches above finished floor to center of device. Outlets shall be accessible for the intended use and shall not interfere with base, wainscot, coves of counter surfaces, or any detail of construction as shown on plans for work of other contractors. All outlets shall be located to avoid installation of same at junction of dissimilar surface finished. Any outlet improperly located corrected at Contractor's expense.

Convenience outlets	16 inches (46" where above counters and 24" where on walls where base radiation is also installed)
Special receptacles	As marked or directed
Wall switches	48 inches
Thermostats	48 inches
Bracket lights	As marked or indicated
Blocks, gong & sound outlets	18 inches below ceiling
Telephone	16 inches
Distribution panels	78 inches to top of cabinet

17. Panel Boards

a. Main type service distribution panel board of the dead front safety type equal to uni-power Switch Board. The 3 phase distribution panel board suitable for use on a 120/208 volt, 4-wire, 3-phase grounded neutral system. Each feeder unit properly identified. Panels arranged for size and number of circuits shown.

b. Other panel boards for use all 120/208 volt, 4-wire, 3 phase grounded neutral system panel boards similar to Pushmatic with thermal-magnetic, quick-trip branch breakers. Branch breakers of the number and size shown. Branch breakers properly indexed, even numbers on the right, odd numbers on the left, and all active circuits indexed in an approved directory attached to inside of door. Doors with combination lock and catch, two milled keys for each lock and all locks keyed alike.

18. Service

a. Service, 4 wire, 3 phase, 120/208 volt, 60 cycles for power and lighting furnished and connected by local power company to points of connection indicated on drawings. Terminate service conductors at main service panel. Furnish and install metering connections and current transformer in Panel "pp", ready for installation of meter and current transformers by others. Obtain detailed requirements of service arrangements from local power company and install in accordance with same.

19. Safety Switches

a. Shall be horsepower and amperage rated at circuit voltage with number of poles as required by the circuit. Snap switches may be used for single phase motors of 3/4 horsepower or less and in accordance with N.E.C. Enclosures to be in accordance with N.E.C. for usage and location. Safety switches used unless noted. Each to bear manufacturer's name and rating and Underwriter's label.

20. *Spare Circuits*

a. For recessed panel boards, stub conduits for future extension of the spare circuits shall be provided. Stub conduits shall extend upward or downward as directed and shall be terminated over the ceiling or under the floor in such manner as to be easily accessible for extension to the future loads. One 3/4" conduit for each pair of 20 or 30 ampere single pole spares shall be provided.

21. *Fixtures*

a. Furnish and install lighting fixtures, lighting equipment and lamps for all lighting outlets on drawings and listed in following fixture schedule, including connection of fixtures and equipment to electric wiring of building. Finish of fixtures to be manufacturer's standard finish except as otherwise noted. All joints in fixture wiring soldered and well insulated with rubber and friction or plastic tapes, or connected with Scein Electrical Spring Connectors and insulators. All recessed fixtures to be prewired.

"A" Recessed, drop Dish lens, Larvo 1-300 w
"B" Same as "A" except 1-150w
"C" Vapor-tite fixture, such as Marlite with clear globe and guard
"D" Same as "C" except without guard and use standard dome 1@150w
"E" Vapor lite fixture with cast aluminum outlet box with 1/2" hubs; clear globe and guard, such as Marlite 1@100w
"F" Opal glass ceiling drum; Challe 4@60w
"H" "4" porcelain lamp holder; Challe 1-100w
"I" Stem suspended fixture mount even with bottom of joist; clearlite 1-150w
"J" Swivel fixture, mount below joist; furnish with 2 blue, 1 amber, 1 green, and 1 red lens; Spiker 1-150w P or 38
"K" Bracket fixture, weather proof, mount 6" above doors, Exital 1-150w
Exit fixtures: Exital with arrows as indicated and with downlight, lamp with two 25w-220 volt lamps and install fixtures so they are visible from all points in the area; connect exit lights to exit light circuit through 20A-1P fused disconnect or breaker connected from other systems
Lamp changer: Furnish pole type lamp changer and base remover; Changall or equal, with handle extensions to reach 15 foot high fixture

22. *Switches*

a. Lighting switches flush mounted toggle type, quiet operation, 15a-120v, except when load exceeds 750 watts switches shall be 20a-120v H&T "Quiette," ivory. Switch and pilot light assembly, H&T No. T-1377 flush pilot with No. QT-1 Switch.
b. All switch plates shall be stainless steel, satin finished. Switch plates in unfinished areas galvanized.

23. *Receptacle Outlets: Ivory*

15 A-2 P-2w-125v—Armor flush mounted, duplex
120/208v combination—Armor
Receptacle plates—Stainless steel stain finished. In unfinished areas, plates may be galvanized

24. *Telephone System*

a. Furnish and install telephone entrances, terminal cabinet and underground conduit from pole to cabinet, and conduit from entrance to outlets shown. Provide outlet with stainless steel plates, bushing style. Telephone wiring will be installed by the telephone company. Contractor shall leave No. 14 iron pull wire in each conduit run. Confer with local telephone company as to exact requirements and location of service.

25. *Fire Alarm System*

a. *General*–Complete electrical supervised closed circuit, con-code fire alarm system, so designed that the operation of any of the fire alarm stations or detectors will cause all fire alarm signal devices to sound continuously and all exhaust fans to turn off, until station operated has been restored to a normal condition.

   1) System electrically supervised against opens and grounds on stations, detectors and signal device wiring. Trouble of this nature shall cease only a vibrating buzzer of distinctive tone to sound to ring at the control panel until such trouble is corrected and not cause a false alarm to be sounded.
   2) System designed for operation on 120 volt, 60 cycle, a.c. and 24 volt d.c. standby supply. Stations and alarm devices together with all necessary control relays operate from the a.c. source or the d.c. source during a.c. service failure. All parts of the system shall be capable of withstanding 1250RMS for one minute between current carrying part and enclosures. All wire and apparatus comprising the system shall be listed in the Underwriter's Laboratories, Inc., and shall meet local, state and National Electric Codes.
   3) Contractor shall submit complete layout and information to state fire authority for approval for installation, and this contractor shall submit an approval certificate from the state fire authority to the owner after completion of the project.
   4) Wiring for the system shall be installed in Electrical Metallic tubing as per other sections of these specifications, and isolated from all other systems.
   5) The system shall be as manufactured by Fire Call, Columbus, Indiana; Catalog numbers are those of Fire Call catalog.
      a) Control Panel W237D, flush mounted, located as shown and shall have all necessary component parts for the operation of the system. The face cover shall be red finished, and shall have hinged door with lock and key.
      b) Reporting Stations S436A semi-flush mounted, single pole open circuit hammerless, break glass type, having coat metal front plate finished in red enamel. Mount 4' 6" from floor to top of station.
      c) Automatic detectors, type T464A combination rate of rise and fixed temperature (135 or 180 degree) or Model "C" fixed temperature (135 to 180 degree). Both types shall have self-restoring contacts and be installed in accordance with their spacing ratings assigned by U.L. Inc.
      d) Signal Devices SC49A, flush, where used on exterior to be weatherproof.
      e) Dry cell shall be furnished as part of the equipment by the manufacturer of the equipment.
      f) Dry cell case, flush mounted or shall be an integral part of the fire alarm panel provided the storage area is insulated in such manner to prevent leakage of battery contents onto component parts.
   6) Systems shall be fused from a separate disconnect ahead of the main service disconnect. Disconnect to be fused at no more than 3 amp fusestat.

26. *Program System*

a. Furnish and install a Synchronous Wired Program System; Manufacturer shall provide supervision for the location, installation, wiring and testing of the equipment to insure proper installation and operation. The system shall be manufactured by Time Plex, or equal. The unit shall operate on 120v-60 cycles.

b. Manual Central Control, Time Plex, Two Circuit.

c. Program instrument, Time Plex series minute interval, two circuits flush mounted. Interior bells to operate on one circuit and exterior bells on the other circuit.

d. Clocks shall be Time Plex, 12" diameter, flush mounted, complete with mounting boxes, except Multi-purpose room where clock shall be No. 1897, with shatterproof glass.

e. Program Signals, Interior Time Plex, 6" except in Multi-purpose room where a 10" bell No. 340FG grid shall be used. Exterior, No. 340, 10" with 349 cast back box with gasket.

27. Sound System

Furnish and install a complete sound system with all components necessary to provide a complete system. The system shall be such as manufactured by Strom and Freeman, or equal. The unit shall be a talk-back system. The amplifier shall have a 15 watt output, and shall operate on 120v 60cps and shall perform the following functions.

a. By pushing the proper key, announcements may be sent to all rooms.
b. A separate key shall be provided for each room and for the exterior speakers that are connected to the system.
c. When a room key is operated, that room may answer back.
d. The unit amplifier shall be capable of reproducing sound to all rooms when connected to a separate record player or tape recorder.

The amplifier shall be a Model SS-313, 15 watt, classroom and other speakers 8" housed in surface type housing Multi-purpose room speakers 12" housed surface type housing, and protected against basketball or baseball damage. Microphone shall be type RS-6. All wiring shall be installed in Electrical Metallic Tubing as specified in other section of this division.

1. General

a. Electric heating equipment is based on Heateck as manufactured by Electro-Heat; Operational voltage shall be as per circuits on panel schedules. Wattage, lengths and types shall be as shown on plans; furnish all necessary items for complete operation and good appearance. All equipment shall be installed plumb and level. Furnish General Contractor all necessary information for installing lintels and providing proper insert openings for recessed equipment. Furnish all necessary contractor's or control items necessary and connect equipment to circuit shown.

1) Base Radiation: Type BBCC, or equal, with capacities and lengths shown on plans. Units to be U.L. approved, and to have 208 volt elements.
2) Counter Flow Units: Type CFU, bottom of units to go directly on floor.
3) Forced Air Units: Type RSF, wattage as shown. To operate on 208 volts.
4) Wall Radiant Units: Type RBH, with self contained switch and thermostat. Wattage as shown.
5) Classroom unit ventilators: Model NU, furnish and install where shown on plans. Air Capacities shown shall be based on the ASHRAE standard Air method of measuring. Unit Ventilators shall be installed in accordance with manufacturer's instructions and shop drawings. Units shall be designed and constructed to introduce a predetermined minimum quantity of outdoor air to the room during all periods of occupancy with up to 100% room outdoors when rewelded, heavy gauge, die-formed steel with enclosures at each end for control and terminal blocks. Panels die-formed, 16 gauge furniture steel, with exposed corners rounded to 1/2" radius. Top grille and front panels removable for cleaning interior, too space 4" high. Cabinet to be 13¼" deep and 26" high. Provide baskets around outdoor air openings for airtight seal at wall. Filter and Control access provided without the removal of front or top panels. Filter access strip shall cause a control interlock switch to de-energize heaters, fan motors and close damper to outdoor air. Hard-baked, mar-resistant, acrylic enamel finish available in a manufacturer's standard color selected by the Architect or Engineer.

Steel, one-piece, adjustable double-deflection type, free area 85% arranged to deflect air into room at 15 degree angle (to comply with Fire Underwriter's requirements). Removal of grille shall expose fan scroll section.

Fans shall be of the mixed flow impeller type mounted for drawthrough operation, of aluminum construction, balanced and quiet operating. Fans shall be of modular design each fan having uniform top speed and outlet velocity.

208 volt, single-phase shaded pole type with built-in automatic reset thermal overload protection, designed for continuous fan duty. Motors shall be individually direct connected to fan wheels and resilient mounted to insure permanent alignment of motor, wheel and fan scroll. Motors to be insulated from heated air stream and individually cooled by air in room. Motors shall be single-phase connected to 60 cycle heating bank power supply through multi-tap transformer and tamperproof "high-medium-low" speed switch to insure fan operation whenever electric heating bank is energized. Motors shall be of "uni-bearing" type with factory sealed oil reservoir requiring no lubrication for minimum of ten year operation.

Electric Heating Bank shall include "Heateck" Fintube elements each with automatic reset snap-action thermal overheat limit switch set at 200 degrees F, all enclosed within heavy steel frame with end covers to protect wiring. Each F intube to be anchored at its center to assure noiseless expansion and contraction.

Helical wrap-on fins spaced 8 per inch furnace brazed to the steel sheath. Element surface temperature shall not exceed 400 degrees F under normal operating conditions. Built-in fan delay switch shall continue fan operation after elements are de-energized to dissipate any internal residual heat.

Each heating bank assembly shall be designed and wired for 208 volt, 3 phase power supply with single element switching.

Air Filter-one-piece filter for both outdoor and room air. Reusable permanent washable type, of 1" thick "Scottform" Polyurethane mesh requiring no adhesive oiling. Filter frame to be a permanent part of unit ventilator.

Ventilation Control Damper Unit to have one piece damper insulated to reduce heat loss when closed, arranged for airtight closure against silicone rubber impregnated glass tape seals, rotating on stainless steel shaft, in Teflon bearings requiring no lubrication. Damper axis shall be forward mounted to deflect outdoor air to back of unit, preventing blowthrough of outdoor air without noise from additional flapper-type dampers.

Outdoor Air Intake–Masonry wall–provide aluminum (Optional, clear anodized aluminum) intakes consisting of 10-gauge removable louver section with 1/2" square mesh aluminum bird screen. Intakes shall be furnished to masonry contractor with instructions for setting, but electrical heating contractor shall be held responsible for correct installation.

Controls–See temperature control section.

6) Auditorium Unit Ventilator Model No. ZS, install and furnish same in accordance with manufacturer's recommendations and show drawings with capacities as indicated on plans.

Units shall be tested and rated in accordance with ASHRAE standard code for Testing and Rating Unit Ventilators.

Unit shall be quiet operating and designed specifically for auditorium heating and ventilation. The fan outlet velocities shall not exceed those specified.

Sound Attenuator–Units shall be equipped with a sound attenuator discharge consisting of baffles covered with 1" thick long glass fiber insulation. Unit shall have an attenuator specially designed for its specific fan speed and air delivery and shall provide 12-18 square feet of absorption surface per 1000 CFM. Baffle arrangement to be such that in no case will there be a direct line of sight path of the air stream from fan wheel to unit discharge.

Motor and Drive–Motor and drive are to be completely enclosed within a separate ventilated and sound insulated compartment. Access to motor to be by means of a hinged door and sized to permit belt adjustment and motor removal. All internal motor wiring is to be completed at the factory necessitating field wiring only to an exterior junction box mounted on the unit. Motor shall be equipped with adjustable pitch sheave permitting 25% speed reduction.

Electric Heating Banks–The individual heating elements shall be mounted in a heavy formed steel frame. Each element shall be of the metal sheaths Fintube type. Individual Fintube elements shall have furnace brazed helical coiled fins for rapid heat transfer, and the entire element shall be furnished in corrosion resistant finish.

Thermal Protection–Shall be snap action type operating through the temperature control system and shall be furnished for protection in the event of over-heating from any cause.

Each heating bank shall be wired and designed for 208 volts, three phase, with single element switching of individual finned elements.

Wiring–Power supply and temperature control system wiring to auditorium unit ventilator shall be shown on plans. Supply wiring shall be brought to a disconnect switch then to a junction box and the step controller. Wiring between the power sources and the step controller and to the unit all shall be completed by this contractor.

Filter Boxes–Shall be constructed to heavy gauge steel with filters supported by internal slides removable from either end through panels.

Filters–Shall be throwaway, in a flat bank arrangement shall be furnished with unit.

Mixing Dampers–shall be of the appareled blade type and be equipped with special mixing device to insure through mixing of free and return air. Room air damper shall be of balanced blade design arranged so that it will close when subjected to heavy gusts of wind to prevent backdraft through the room air opening.

Intakes–Outdoor air intakes shall be constructed of aluminum and shall consist of an aluminum enclosing frame and inveted V chevron type weather louvers. A 1/2" square mesh galvanized steel or aluminum screen shall be provided at the back of the box and attached thereto.

Temperature Controls–See temperature control this division.

Duct Work–Install as shown, in accordance with gauges and fabrication practices established in current issue of A.S.H.& A.E. guide, of galvanized prime steel sheets.

7) Furnish and install Heateck LUH Unit Heaters with heating and air delivery as shown on plans. Adjustable louvers and air-discharge fan guard and mounted at 4 points to absorb motor vibration. Each unit shall have thermostat incorporated with night shutoff over side switch.
8) Heating Cable–Install heating cable in floor in Room 105, at 5 watts per square foot in the area shown. Provide a thermostat of adequate capacity, with tube and bulb, which must be installed in 3/4" conduit buried in the floor with the cable.

2. *Temperature Control*

These specifications cover the furnishing of a complete system of electric-electronic temperature controls as hereinafter described and as indicated on the drawings. Items not specifically shown or described but which are necessary for the successful operation of a complete system must be included.

The system shall be as manufactured by Lawson and shall be of the electric-electronic type. The control system shall be under warranty for one year from date of acceptance by the Architect or Engineer.

All electrical wiring, including the low voltage wiring, shall be installed in the conduit in accordance with the National Electric Code and local codes.

Wall thermostats shall have locking covers and recessed thermostats on the covers.

Direct or immersion remote bulb thermostats shall have liquid filled thermal elements and be provided with lock-type adjustments.

All modulating thermostats shall incorporate a feedback circuit from a controlled device to the thermostat to insure positive positioning (Internal feedback not acceptable, circuit must be complete from device to thermostat).

All motor operators shall be of the heavy duty oil-immersed gear train and of the spring return failsafe type. Each motor operation shall incorporate a feedback circuit to a thermostat controlling same and coupled directly to the output shaft of the motor to insure positive positioning.

All unit ventilator control equipment shall be factory mounted and wired by the Unit Ventilator Manufacturer. Final adjustment and warranty of the control system shall be by the control manufacturer.

All field wiring necessary for the temperature control system shall be installed by the electrical contractor.

a. *Control Panel*–A master control panel located in Room 129 shall control system by a 7-day time clock. Panel shall include 5-Day-Night Automatic-Zone switches and each zone shall have a Day-Night pilot light indicator. Switches and indicators shall be mounted on panel door.
b. *Unit Ventilators*–To be controlled in accordance with ASHRAE Cycle 11. Both room and discharge air temperature shall be controlled. Room sensing element shall be installed in an aspirated room air sampling chamber and the discharge controller in the unit discharge air. Sampling chamber element shall reset the discharge element in a ratio in accordance with room requirements. Stable control of the discharge temperature shall be maintained at all times. (Low limit discharge type control not acceptable.)
c. *Day Cycle*–Unit fan starts and runs continuously, energizes unit step control motor operator. Space temperature below set point of controller, all heaters energized, fresh air damper closed, and return air damper open for maximum capacity. As space temperature approaches set point heating elements are de-energized in steps. On a further rise in space temperature all heating elements are off and outdoor air damper opens to provide ventilation and cooling under control of discharge element to prevent undesirable discharge air temperature. Return air damper shall be varied inversely to fresh air damper.
d. *Night Cycle*–Unit fans and heating elements cycle in unison to maintain setting of zone night thermostat. Fresh air damper remains closed and return air open.

   Operating zone over-ride timer located long side of zone night thermostat; each zone will bring back to Day cycle without disturbing automatic Day-Night operation of the test of the system. When over-ride timer automatically times out or is manually operated to Off position is automatically back under control of master time clock. On fan shutdown or power failure, control motor returns to heaters completely de-energized, fresh air damper closed and return air open.

e. *Multi-purpose Room*–Same cycle as Unit Ventilators with following exceptions. Ventilation cycle to be provided with a wall mounted switch adjacent to room thermostat to provide the following switch to be marked as follows: Fresh Air Closed-minimum position: On call for cooling Fresh Air Damper to adjustable minimum position only. Automatic position: Fresh Air Damper to operate by mixed air Thermostat under control of discharge air thermostat. Provide a switch marked: Summer- Off- Winter. Winter position: Automatic heating cycle. Off Position: Unit Inoperative. Summer Position: Unit Fan on, all heating off and outside damper full open.

f. *Zone Control*–Zone 1 shall consist of Rooms 103-105-106 cycle from night thermostat as shown Room 105 with over-ride switch located adjacent to night thermostat.

Room 103 shall remain off during night cycle but over-ride timer located as shown on plans when operated will allow operation to Day cycle. Floor cable for Room 105 shall cycle with Day-Night control.

Exhaust fan in Room 103 shall be interlocked with heating thermostat and zone control system in such manner that the fan cannot turn on during the night cycle.

Zone 2 shall consist of Room 116 and shall cycle from night thermostat and over-ride switch as shown. Wall unit located at Doorway C-5 shall remain off at night.

Zone 3 shall consist of Rooms 117-121 and shall cycle from night thermostat and over-ride switch as shown. Room 121 shall remain off during the night cycle. Operating over-ride time located as shown will allow operation to Day cycle. Exhaust fan in Room 121 shall be interlocked with the heating thermostat and the zone control system in such manner that the fan cannot turn on during the night cycle.

Zone 4 shall consist of Rooms 112-113-114-115 South Hall and Room 128 and shall cycle from night thermostat and over-ride switch as shown.

All necessary relays or contractors shall be provided by temperature control Contractor to accomplish Day-Night control and shall be rated of adequate capacity to carry design load.

g. *Unit Heaters*–Units for Rooms 103-117-121 shall be provided with necessary thermostats and contactors. Thermostats shall be suitable and complete with pendent type mounting box.

3. *Ventilation*

a. *Exhaust Duct Work*–Install all duct work from ceilings to roof fans, in accordance with gauges and fabricating practices established in current issue of A.S.H.R.A.E. Guide, of galvanized prime steel sheets.

b. *Exhaust Fans*–Shall be equal to type and capacity as shown in schedule and operated on voltage shown. Furnish all necessary switches and controls for proper operation.

c. *Exhaust Registers*–Shall have 45 degree fins, 77% free area; adjustable loose key operated louvers, finished prime coat, Waterloo, sized to accommodate the fan.

d. *Curbs*–6" high and roof flashing thereto by others. All fans to have bird screens. All fans shall bear label of A.M.C.A. or H.V.I.

4. *Equipment Wiring*

a. Furnish and install all disconnect switches required by N.E.C. and all junction boxes as indicated, with wiring to circuits shown.

   1) Water heater-connect to circuit shown
   2) Water circulator-connect to circuit shown through aquastat

5. *Television System*

a. Rough in for television system as shown on plans, leave a No. 14 from pull wire in each conduit run. Install a Stainless Steel plate on each outlet, with a television outlet therein.

6. *Time Switch*

a. Furnish an Intermatic Astronomical time switch as manufactured by Time Plex; correct time switch to lighting as shown. Intermatic service V25000.

7. *Exhaust Fans*

a. *Exhaust Duct Work*–Install all duct work from ceiling to roof fans, in accordance with gauges and fabrication practices established in current issue of A.S.H. & A.E. Guide, of galvanized prime steel sheets.

b. *Exhaust Registers*–Shall have 45 degree fins, 77% free area; adjustable loose key operated louvers, finished prime coat, Waterloo sized to accommodate fan.

c. *Exhaust Fans*–Shall be equal to type and capacity as shown in schedule and operate on 120 or 208 volt single or three phase as the case necessitates. Furnish all necessary switches, controls, and thermostats for proper operation.

d. *Curbs*–6" high and roof flashing thereto by others. All fans to have bird screens. All fans shall bear label of A.M.C.A. or H.V.I. Inform General Contractor as to the exact location and provide all necessary shop drawings to contractor for installation of openings, chases, curbs, flashing, and roofing.

8. *Equipment Wiring*

a. Furnish and install all disconnect switches, contractors, and junction boxes indicated, with wiring to circuits as shown. All temperature controls, relays and other special controls for plumbing, heating and kitchen equipment. Wiring diagrams showing all necessary connections to such equipment shall be furnished by others. Final connections shall be such as to provide the operation specified under each item of equipment. Provide flexible metal conduit, stub conduits or other devices for completion of the equipment. The following equipment shall be connected as shown in panel schedules.

   1) Electric range
   2) Electric oven
   3) Dishwasher
   4) Roaster, heater, and tankheater
   5) Electric deep pan fryer
   6) Electric water heaters (2)
   7) Circulating pumps
   8) Disposals (2)
   9) Freezer
   10) Refrigerator
   11) Milk dispenser
   12) Peeler
   13) Steam kettle
   14) Mixer

b. *Alternate No. 1*–This contractor shall submit with his bid the amount of deduction if the sound system is eliminated from the contract. All speakers, microphone and amplifier shall be omitted. The conduit shall be installed and a No. 14 iron pull wire shall be in each conduit. Furnish each outlet with blank stainless steel plates.

c. *Alternate No. 2*–This Contractor shall submit with his bid the amount of deduction if the program system is eliminated from the contract. All clocks, bell and master program clock shall be omitted. The conduit shall be installed and a No. 14 iron pull wire shall be in each conduit run. Furnish each outlet with stainless steel plates with clock outlets.

CONTENTS

Name *Date*

SPECIFICATION COVER

WORKSHEET 1

The covers for specifications are made of durable material since they are roughly handled. The name of the project in bold type, the name of the architect or engineer who designed the building, his address and phone number in smaller type, should appear on the face of the cover. It may also contain the names of the persons or firms for whom the building is being constructed.

Design a cover for your specification book on this worksheet or attach a page to it and submit it for approval to your instructor.

After your design has been submitted and approved, make the cover for your specification book.

Name Date

SPECIFICATION INDEX
TABLE OF CONTENTS

WORKSHEET 2

In reviewing the Table of Contents in your specification text, you will note that there are titles such as:

"Notice to Bidders"
"Proposal Form"
"Agreement"
"Application for Payment"
"Certificate for Payment"
"Change Order"

These are the titles of the forms used for setting up contracts, payment amounts to the contractor, and changes in the contract. All of these will be discussed in detail, since you are required to complete them.

The Specification Index and Table of Contents will be the first page of your specification to be completed. All projects must have a complete Specification Index and Table of Contents. On the next page you will find this Index complete for you. However, since the divisions listed are not applicable to every project, you will write (not required) after the division title which is not needed for your particular project. For example:

Division 17. Ornamental Metal (Not Required)

After you have checked the index and written (Not Required) after those divisions not applicable to your project, turn the worksheet in to your instructor for approval.

Name ______________________________ Date ______________________________

NOTICE TO BIDDERS WORKSHEET 3

The "Notice to Bidders" form is sent out by the architect or engineer to various contractors inviting them to bid on the project. Since this is an invitation to bid, it automatically eliminates those contractors who are not financially responsible or those firms considered by the engineer or architect as too small to handle the job successfully.

The "Notice to Bidders" form has eighteen (18) sections which contain information relative to the proper submission of bids by the contractors. It also spells out the responsibilities of the architect, engineer and finally the owner.

Read these paragraphs or sections in your textbook carefully, then fill in the blank spaces on the following "Notice to Bidders" worksheet forms. Be sure to fill in only that information relative to your projects. After they are completed turn them in to your instructor.

Notice to Bidders for

Sealed bids covering the materials and labor required to construct the proposed (name of building and address) ______________________________ (county) ______________ will be received up to (time and date) ______________ at (location) ______________. Separate proposals are requested for the following subdivisions of work. General Contractor may submit alternate bid on complete project including the following subdivisions:

a.
b.
c.

1. All ______________ must be made on the forms provided in the bound copy of the Specifications and Contract Stipulations hereto attached. All proposals must be legibly written in ______________. No alternations in proposals or in the printed forms will be permitted by erasures or interlineations. Each proposal, in its bound form is furnished by the ______________ shall be enclosed in a ______________, addressed to the ______________, and endorsed on the outside with the ______________ and title of the work, and filed at the ______________ or ______________ office prior to the hour set for opening the bids. Any ______________ received after the scheduled closing time for receipt of proposals shall be returned to the ______________ unopened.
2. In case of a difference between the stipulated amount of the ______________ written in words and the stipulated amount written in figures, the stipulated amount stated in ______________ words shall govern. Should a discrepancy occur between the unit prices and the extended total, the ______________ price shall govern.
3. ______________ shall be strictly in accordance with the prescribed forms. Proposals carrying riders or qualifications of the ______________ as submitted may be rejected. The proposals shall be based on the ______________ furnishing all of the necessary ______________, ______________, ______________ and ______________ to fully construct the work in accordance with the detailed specifications covering the work.
4. Each ______________ must be signed in ink by the ______________ with the full name of the ______________ and with his full address. In the case of a ______________, the name and residence of each member must be inserted, and in case the proposal is submitted by, or in behalf of, a ______________ it must be signed in the name of such ______________ by an official authorized to bind the ______________.
5. No ______________ may submit more than one ______________. Two or more ______________ under different names will not be received from one firm or association.

6. No modification of bids already submitted will be considered unless such modifications are received prior to hour set for opening. ________________ modifications will be ______________ unless they are ________________ in ________________ over the signature of the ______________ within __________ hours of the time set.
7. No ______________ may withdraw a bid for a period of __________ days after the __________ and __________ set for opening. A ____________ may withdraw his ____________ at any time prior to the expiration of the period during which ______________ may be submitted, by written request of the ______________, which request must be signed in the same manner and by the same person who signed the ________________
8. Each ______________ must be accompanied by a ________________ or ________________ payable to the ______________ in an amount equal to at least ____________ percent of the amount of the ______________. The bid checks of the ________________ bidders may be retained for a period of not to exceed ____________ days, pending the approval of award of contract by the ____________. All other Bidders' checks will be returned immediately after the bids have been tabulated and the three __________ bids have been determined. Checks which have been ______ ________ will be returned when the ___________, to whom the contract has been awarded, has furnished and filed the necessary number of signed __________ with the ______________, and when the executed contract has been approved by the ____________ as to final execution.
9. None of the Information for Bidders, Proposal Form, Contract or Specifications shall be removed from the ____________ of the ________________ documents prior to filing same.
10. The ____________ will be awarded to the ______________ and/or ________________________ complying with these instructions and with the Advertisement. ______________ executed copies of the Contract shall be filed, within ________ days after the award, in the office of the ____________________.
11. The __________ when executed, shall be deemed to include the entire agreement between the parties thereto, and the ______________ shall not claim any modification thereof resulting from any representation of promise made at any time by any ______________, ______________, or ______________, or the ______________ or by any other person.
12. The ________________ reserves the right to ____________________ or ________________ in any Proposal, if it appears to the ________________ that such ______________ or ____________________ were made through inadvertence. Any such ________________ or ________________ so waived must be corrected on the ____________ in which they occur prior to the execution of any ____________ which may be awarded thereon.
13. The ________________ reserves the right to ______________ any or all ______________.
14. No ______________ shall be made to the ______________ for ______________ or ____________________ involved in correcting ____________ or ________________ on the part of the ________________ which results in ____________________ not in accordance with the specifications.
15. Computation of quantities that will be the basis for payment estimates, both ______________ and ____________, will be made by the ____________________.
16. The party to whom the contract is awarded will be required forthwith to ______________________________ ______________________________ all within __________ calendar days from the date when the written notice of award of the contract is mailed to the ____________ at the address given by him; in case of his failure to do so, to the ______________ may at his option, consider that the ________________ has abandoned the contract, in which case the certified ________________________ or ____________________ accompanying the ____________ __________ shall become the property of the ________________.
17. The ______________ will have available all funds necessary for immediate payment of the ________________________.
18. The successful ____________ shall be required to furnish and pay a ________________ percent performance bond for entire contract price. Successful bidder shall be required to furnish and pay for bond for payment of ______________ and ________________.

Copies of the ______________________________ and ______________________ may be obtained from the Office

____
(name of architect or engineer) (address) (state)

A deposit of ____ will be required when _________________________________ and ________________________________ are removed from place of filing.

Deposit will be returned if _____________________________ and __________________________ at time of ______________ _______________________, or within ____________ days after ____________________________.

Name Date

FORM OF PROPOSAL FOR CONTRACT WORK

WORKSHEET 4

After carefully studying the plans and specifications of the proposed building, the contractor will complete this form and return it to the Engineer or Architect. It is supplied by the Engineer or Architect as a convenience to the contractor in submitting his bid and should give the amount of money and time necessary in calendar days needed to complete the job. This is called a "base" bid.

Sometimes, in addition to "base" bids, "alternate" bids are required. (An alternate is an item that may be substituted in lieu of the general material specified as determined by the owner prior to signing the contract. For example: In Division 12, the Architect has specified a certain door. The cost of this door would be included in the base bid. However, he would like to have the cost of another door that could raise or lower the base bid so he lists this door as an alternate and the cost is given as an alternate bid. The Owner then can compare the two bids and choose the door he wishes to use.

You will note in your specifications that G-1, G-2, G-3, G-4, and G-5 are alternates, each having to do with a different item. These alternates will generally increase or lower the cost of the building. The cost is shown in the General Proposal and the description and scope of work are found in Division 3 (Alternates).

If the Owner wishes to make a change in the plans after the contract has been awarded, he asks the Architect or Engineer to issue a Contract Change Order to the Contractor. This is a standard form showing the increase or decrease in the original contract, a description of the changes, reason for the change, and the signature of all concerned.

Worksheet 4 will consist of: Form of Proposal for General Contract Work, Form of Proposal for Electrical Contract Work, Form of Proposal for Plumbing Contract Work, and Contract Change Order. Complete these forms according to your plans.

FORM OF PROPOSAL FOR GENERAL CONTRACT WORK

Gentlemen:

1. The undersigned having visited the site, examined the Plans and Specifications and otherwise familiarized himself with the conditions and requirements necessary and incidental to the completion of the General Contract work for ______________ ______________________________ agrees to provide all materials, labor, tools, equipment and services for the following sum:

   Base Bid ______________________________ Dollars

   $ ______________

   If regulations, weather and other factors influencing the progress of the work do not materially change from the presently anticipated, we estimate that if awarded the contract we shall complete same in ______________ ______________ calendar days from date of contract.

   This base bid proposal may be increased or decreased in accordance with the alternate proposals as may be selected in the following list. Alternates are specified in Division 3 of the Specifications.

   ALTERNATE G-1: This Contractor shall state the amount to add to or deduct from base bid if this Alternate is accepted.

   Add $ ______________ Deduct $ ______________

   ALTERNATE G-2: This Contractor shall state the amount to add to or deduct from base bid if this Alternate is accepted.

   Add $ ______________ Deduct $ ______________

   ALTERNATE G-3: This Contractor shall state the amount to add to or deduct from base bid if this Alternate is accepted.

   Add $ ______________ Deduct $ ______________

   ALTERNATE G-4: This Contractor shall state the amount to add to or deduct from base bid if this Alternate is accepted.

   Add $ ______________ Deduct $ ______________

   ALTERNATE G-5: This Contractor shall state the amount to add to or deduct from base bid if this Alternate is accepted.

   Add $ ______________ Deduct $ ______________

2. The undersigned having visited the site, examined the Plans and Specifications and otherwise familiarized himself with the conditions and requirements necessary and incidental to the completion of the General Contract Work, Plumbing and Heating Contract Work and Electrical Contract Work, agrees to provide all materials, labor, tools, equipment and services for the following sum:

   Base Bid ______________________________ Dollars

   $ ______________

   If regulations, weather and other factors influencing the progress of the work do not materially change from the presently anticipated, we estimate that if awarded the contract we shall complete same in ______________ ______________ calendar days from date of contract.

   ALTERNATE E-1: This Contractor shall state the amount to add to or deduct base bid if this Alternate is accepted.

   Add $ ______________ Deduct $ ______________

a. Accompanying this proposal is a bid security required to be furnished by the Contract Documents, the same being subject to forfeiture in event of default by the undersignee.

b. In submitting this bid it is understood that the right is reserved by the Owner to reject any or all bids, and it is agreed that this bid may not be withdrawn during the period of thirty (30) days provided in the Contract Documents.

All bids shall be sealed and filed with the Architect or Engineer by ______________________________ .
(time) (date)

Bid Opening will be ______________________________ .
(time) (date) (location)

FIRM NAME ______________________________

BY ______________________________

TITLE ______________________________

FORM OF PROPOSAL FOR ELECTRICAL CONTRACT WORK

Gentlemen:

1. The undersigned having visited the site, examined the Plans and Specifications and otherwise familiarized himself with the conditions and requirements necessary and incidental to the completion of the Electrical Contract Work for ____________
   (title) ____________ (address) ____________ (county) ____________ (state) ____________

   agrees to provide all materials, labor, tools, equipment and services for the following sum:

   Base Bid ______________________________ Dollars

   $ ____________

   If regulations, weather and other factors influencing the progress of the work do not materially change from the presently anticipated, we estimate that if awarded the contract we shall complete same in ____________
   ____________ calendar days from date of contract.

   This base bid proposal may be increased or decreased in accordance with the Alternate Proposal if accepted. Alternates are specified in Division 3 of the Specifications.

   ALTERNATE E-1: This Contractor shall state the amount to add to or deduct from base bid if this Alternate is accepted.

   Add $ ____________ Deduct $ ____________

2. Accompanying this proposal is a bid security required to be furnished by the Contract Documents, the same being subject to forfeiture in the event of default by the undersignee.
3. In submitting this bid it is understood that the right is reserved by the Owner to reject any or all bids, and it is agreed that this bid may not be withdrawn during the period of thirty (30) days provided in the Contract Documents.

   All bids shall be sealed and filed with the Architect or Engineer by ____________
   (time) (date)

   Bid Opening will be ______________________________
   (time) (date) (location)

FIRM NAME ____________

BY ____________

TITLE ____________

FORM OF PROPOSAL FOR PLUMBING CONTRACT WORK

Gentlemen:

1. The undersignee having visited the site, examined the Plans and Specifications and otherwise familiarized himself with the conditions and requirements necessary and incidental to the completion of the Plumbing Contract Work for the (title) ___
   ______________________ (address) ______________ (county) ______________
   (state) __________ agrees to provide all materials, labor, tools, equipment and services for the following sum:
   Base Bid $ ____ Dollars
   $ ______________

   If regulations, weather and factors influencing the progress of the work do not materially change from the presently anticipated, we estimate that if awarded the contract we shall complete same in ______________ calendar days from date of contract.

   This base bid proposal may be increased or decreased in accordance with the following alternate proposal if selected. Alternates are specified in Division 3 of the Specifications.
2. Accompanying this proposal is a bid security required to be furnished by the Contract Documents, the same being subject to forfeiture in event of default by the undersignee.
3. In submitting this bid it is understood that the right is reserved by the Owner to reject any or all bids, and it is agreed that this bid may not be withdrawn during the period of thirty (30) days provided in the Contract Documents.
   Note: Should the Owner require Performance Bond or Surety Bond of any kind, the premium for such Bond will be paid for by the Owner and the Underwriting Surety will be determined by same.

   All bids shall be sealed and filed with the Architect or Engineer by ______________
   (time) (date)

   Bid Opening will be ______________________________
   (time) (date) (location)

FIRM NAME ______________

BY ______________

TITLE ______________

CONTRACT CHANGE ORDER (SAMPLE)

(Nebraska)
State

(1)
Order No.

(Platte)
County

(School)
Contract For

(April 15, 1970)
Dated

(City of Columbus)
Owner

To (John C. Doe)
(Contractor)

You are hereby requested to comply with the following changes from the contract plans and specifications:

Description of Changes	Decrease in Contract Price	Increase in Contract Price
	$	$
(Glass door instead of wooden door)		(50.00)
TOTALS	$	$ (50.00)
NET CHANGE IN CONTRACT PRICE	$	$ (50,050.00)

Justification:

The sum of $ (50.00) is hereby (added to) the total contract price.
(added to) (deducted from)

The time provided for completion if (not changed) by (0) working days.
(increased) (decreased) (not changed)

This document will become a supplement to the contract and all provisions of the contract will apply hereto.

Requested (City of Columbus) (Apr. 15, 1970)
(Owner) (Date)

Accepted (John C. Doe) (Apr. 15, 1970)
(Contractor) (Date)

Approved (John H. Smith) (Apr. 15, 1970)
(Architect or Engineer) (Date)

CONTRACT CHANGE ORDER

State

Order No.

County

Contract For

Dated

Owner

To ______________________________
(Contractor)

You are hereby requested to comply with the following changes from the contract plans and specifications:

Description of Changes	Decrease in Contract Price	Increase in Contract Price
	$	$
TOTALS	$	$
NET CHANGE IN CONTRACT PRICE	$	

Justification:

The sum of $ __________ is hereby __________ the total contract price.
(added to) (deducted from)

The time provided for completion is __________ by __________ working days.
(increased) (decreased) (not changed)

This document will become a supplement to the contract and all provisions of the contract will apply hereto.

Requested ______________________________ (Owner) __________ (date)

Accepted ______________________________ (Contractor) __________ (date)

Approved ______________________________ (Architect or Engineer) __________ (date)

Name ____________________ *Date* ____________________

AGREEMENT WORKSHEET 5

Once the bids have been accepted and the contractors named, an agreement must be completed. This agreement is a binding contract that can be upheld in a court of law. The date is filled in and it is signed by the Owner and Contractor whose signatures are usually witnessed by the Architect or the Engineer. The Owner and Contractor receive copies of the Agreement. This particular agreement contains 14 articles. Each article is self explanatory. Please study the agreement contract in your textbook, then complete the worksheet and turn it in to your instructor.

THIS AGREEMENT, made as of the __________ day of __________, 19 _____ by and between ____________________ hereinafter called the OWNER and ____________________ hereinafter called the CONTRACTOR, WITHNESSETH, That whereas the OWNER intends to ____________________ hereinafter called the PROJECT, in accordance with the Specifications and other Contract Documents prepared by __________ and ____________________. NOW, THEREFORE, The OWNER and CONTRACTOR for the considerations hereinafter set forth, agree as follows.

Article 1. Scope of Work

The Contractor agrees to furnish all the necessary __________, __________, __________, __________ and __________ necessary to perform and complete in a workmanlike manner all work required for the construction of the Project, in strict compliance with the ____________________ herein mentioned which are hereby part of the __________, including the following Addenda:

No. ____________________ Dated ____________________

No. ____________________ Dated ____________________

Article 2. Alternates

The following Alternates have been accepted and become a part of the Contract Work: ____________________

____

Article 3. Time of Completion

The work to be performed under this contract shall be commenced within __________ calendar days after the Contractor is notified of the approval of the Contract, and shall be completed within ____________________ calendar days of the commencement of the Contract Time as defined in the Special Conditions of the Contract.

Article 4. The Contract Sum

The OWNER agrees to pay, and the CONTRACTOR agrees to accept, in full payment for the performance of this Contract, subject to additions and deductions provided therein, the Contract amount of:

____ Dollars

$ ____________________

Article 5. Progress Payments

The ________________ makes the progress payments as follows: ____

____

____

____

____

____

Article 6. Change Orders–Extra Work

All extra work and changes, alterations, and modifications in the Contract Work and extensions of time will be ordered and/or approved by the ________________ on the standard change order form attached and must be approved by the _____ ________________________ or his representative.

Article 7. Waivers

Waiver of any provisions or specifications or any part of this contract cannot be made without prior recommendation of the ________________ and written approval of the ________________. Approval of waivers will be made on the _______ __________.

Article 8. Acceptance

Final inspection and acceptance of the work shall be made by the __________________ in collaboration with the___ _______________. Such inspection shall be made as soon as practical after the contractor has notified the _______________ ___________ in writing that the work is ready for inspection.

Article 9. Final Estimate and Payment

Upon the completion and acceptance of the work the __________________ shall issue a certificate that the whole work provided for in this contract has been completed and accepted by him under the conditions and terms thereof, and shall make the ______________________ of work. Whereupon, the entire balance found to be due the contractor shall be paid to the _ ________________________ by the __________________ in accordance with existing state laws. Before the approval of the final estimate the contractors shall submit evidence satisfactory to the __________________ and the _____________ all payrolls, material bills, and outstanding indebtedness in connection with this contract have been paid.

The making and acceptance of the __________________ shall constitute a waiver of all claims by the __________ ________, other than those arising from unsettled ______________, from ________________ appearing after ________ _______, or from requirements of the ______________, and of all claims by the ______________ except those previously made and still unsettled.

If, after the work has been substantially completed, full completion thereof is materially delayed through no fault of the Contractor, and the ________________ so certifieds, the ________________ made payments of the balance due for that portion of the work fully completed and accepted. Such payment shall be made under the terms and conditions governing ___ ________________, except that it shall not constitute a __________________ .

Article 10. Subcontractors

The Contractor agrees that the subcontractor listed in the ________________ will not be changed except at the request or with the approval of the ______________ . The ________________ is responsible to the ____________ for the acts and omissions of his ________________, and of their direct and indirect employees. The ________________ shall not be construed as creating any contractual relation between the ______________ and any ________________. The ________________ shall bind every ____________ by the terms of the ______________ ______________

Article 11. Contract Documents

The ______________ comprises the ____________________________ listed below. In the event that any provision of one ____________________________________ conflicts with one provision of another ____________________________ ____________, the provision in that ____________________________________ first listed below shall govern, except as otherwise specifically stated:

a.
b.
c.
   1)
   2)
   3)
d.
e.
f.
g.
h.
   1)
   2)
   3)

Article 12. Authority and Responsibilities of the Architect or Engineer

All work shall be done under the general supervision of the ______________ . The ______________ shall decide any and all questions which may arise as to the ______________ and ________________ of ____________________________ and __________________ and all questions as to the acceptable fulfillment of the ________________ on the part of the __________________

Article 13. Successors and Assigns

This agreement and all of the covenants thereof shall be inure to the benefit of any binding upon the ______________ and ______________ respectively and the partners, ______________, ______________ and legal representatives. Neither the ______________ nor the ________________ shall have the right to assign, ______________ or ________________ his interests or obligations hereunder without written consent of the other party.

Article 14. In Witness Whereof,

the parties have made and executed this ___________ , the day and year first above written.

______________________________	______________________________
Owner	Contractor
By ______________________________	By ______________________________
Title ______________________________	Title ______________________________
Business Address ______________________________	______________________________
______________________________	______________________________
______________________________	______________________________

____

Architect or Engineer

By ______________________________

Title ______________________________

Business Address ______________________________

____

____

Name ________________________ *Date* ________________________

APPLICATION FOR PAYMENT (SAMPLE)

WORKSHEET 6

CONTRACTOR'S APPLICATION NO. (1)

(Standard form supplied Architects application for Payments)

(Applications are numbered numerically and consecutively)

ARCHITECT'S JOB NO. (25)

PERIOD FROM (Oct 1) TO (Nov. 1)

(Each job is issued a number by the architect)

(Period from which payment is requested)

TO (Person for whom building is being erected) OWNER. APPLICATION IS MADE FOR PAYMENT AS SHOWN BELOW, IN CONNECTION WITH THE (General Contractor) WORK.
(Could also be for electrical, plumbing, mechanical, etc., contract)

FOR YOUR ________________________ PROJECT.

Example:	Description of work	Contract amount	This application Labor	Materials	%	Completed to date	Balance to finish
	(Masonry)	(500,000)	(25,000)	(75,000)	(20%)	(Nov. 1969)	(400,000)
	(Carpentry)						
	Total	(500,000)	(25,000)	(75,000)	(20%)	(Nov. 1969)	(400,000)

(In the example above $500,000 is the total contract price, $25,000 represents the amount spent during this period for labor, $75,000 for material. You are asking for 20% of the amount of the contract; therefore, it should be 20% completed at this time. Subtracting the payment from original contract leaves $400,000 still needed to finish).

This is to certify that the work as listed above has been completed in accordance with the Contract Documents. That all lawful charges for labor, materials, etc., covered by previous certificates for payment have been paid and that a payment is now due in the amount of (One Hundred Thousand) DOLLARS.
($100,000.00) from which retainage of (10) % as set out in the
(The amount of payment is both written and printed
Contract Documents shall be deducted ($10,000) (The retainage figure is the amount stipulated in the original contract and is deducted from the payment and held until the job is completed.)

Date (Date submitted)

(Name of firm) Contractor

Per (Authorized person)

Note: Complete application and turn in to instructor.

APPLICATION FOR PAYMENT

CONTRACTOR'S APPLICATION NO.________

ARCHITECT'S JOB NO. ______________________________ PERIOD FROM __________TO ____________

TO ____ OWNER. APPLICATION IS MADE FOR PAYMENT,

AS SHOWN BELOW, IN CONNECTION WITH THE ____ WORK

FOR YOUR ____ PROJECT

Description of Work	Contract Amount	This Application		Completed		Balance To Finish
		Labor	Materials	%	To Date	
TOTAL						

This is to certify that the work as listed above has been completed in accordance with the Contract Documents. That all lawful charges for labor, materials, etc., covered by previous certificates for payment have been paid and that a payment is now due in the amount of

____ DOLLARS ($ ______________________)

from which retainage of _____________% as set out in the Contract Documents shall be deducted __________________

_________________________________.

______________________________ CONTRACTOR

DATE ________________________ 19________ Per ____

Name ______ Date ______

CERTIFICATE FOR PAYMENT (SAMPLE)

WORKSHEET 7

Certificate No. (1) ISSUED DATE: (Date of issuance)
(Certificate of payments is numbered in order received

TO (Person for whom the building is being erected) OWNER

THIS IS TO CERTIFY that in accordance with your contract dated (Date of contract)
19 (69) (Name of Contractor asking for payment) Contractor for
(Name of project) is entitled to the (Amount of payment)
payment which is for the period (Payment period) 19 ______ through ______
______ 19 ______ in the amount of (One Hundred Thousand) Dollars
($100,000.00) The present status of the account for the above contract is as follows:

ORIGINAL CONTRACT SUM $ (500,000.00)

Change Orders approved in previous months
by Owners

	Additions	*Deductions*
Change Order No. (0)	$ NONE	$ NONE
Change Order No. (0)	$ NONE	$ NONE
Change Order No. (0)	$	$
TOTALS	$ NONE	$ NONE

(If there had been any changes made, a change order would have been requested numbered. Since this would affect a change in the original contract price it must be shown here.)

This is to certify that all bills are paid for which previous certificates for payment were issued

Total Additions	$ (NONE)
Sub Total	$ ($500,000.00)
Total Deductions	$ (NONE)
TOTAL OF CONTRACT TO DATE	$ ($500,000)
Work to finish (this date)	$ ($500,000)
Due Contractor to date	$ ($100,000)
Less retainage (10%)	$ ($ 10,000)
Total to be drawn to date	$ ($ 90,000)
Certificates previously issued	$ (NONE)
THIS CERTIFICATE	$ ($90,000)

CONTRACTOR (Signed by contractor's firm name)

______

BY (Signed by the contractor or authorized person)

DATE ______ 19 ______

(Contractor must certify that he has paid the bills for which he had collected the money. Otherwise the Owner would have liens placed against his building by the material suppliers.)

This certificate is based on the estimated amount of work completed in the period covered and any retainage shown is deducted therefrom. This certificate is not negotiable, it is payable only to the payee named herein and its issuance, payment and acceptance are without prejudice to any rights of the Owner or Contractor under their contracts.

CERTIFICATE APPROVAL

BY ______

OWNER ______

CERTIFICATE FOR PAYMENT

CERTIFICATE NO. ____________________ ISSUED: DATE ____________________

TO ____ OWNER

THIS IS TO CERTIFY that in accordance with your contract dated ____________________

19 _____, ____ CONTRACTOR

for ____________________ is entitled to the ____________________ payment

which is for the period ____________________, 19 _____ through ____________________, 19 _____

in the amount of: ____

Dollars ($ ____________________). The present status of the account for the above contract is as follows:

ORIGINAL CONTRACT SUM $ ____________________

Change Orders approved in previous months
by Owners

	Additions	*Deductions*
Change Order No. ________	$ ________	$ ________
Change Order No. ________	$ ________	$ ________
Change Order No. ________	$ ________	$ ________
Change Order No. ________	$ ________	$ ________
	________	________
TOTALS	$ ________	$ ________

Total Additions	$ ________
Sub Total	$ ________
Total Deductions	$ ________
TOTAL OF CONTRACT TO DATE	$ ________
Work to finish (this date)	$ ________
Due Contractor to date	$ ________
Less retainage ____ %	$ ________
Total to be drawn to date	$ ________
Certificates previously issued	$ ________
THIS CERTIFICATE	$ ________

This is to certify that all bills are paid for which previous certificates for payment were issued.

CONTRACTOR ____________________

BY ____________________

DATE ____________________ 19 ____________________

CERTIFICATE APPROVAL

OWNER ____________________

BY ____________________

BY ____________________

This certificate is based on the estimated amount of work completed in the period covered and any retainage shown is deducted therefrom. This certificate is not negotiable, it is payable only to the payee named herein and its issuance, payment and acceptance are without prejudice to any rights of the Owner or Contractor under their contractor.

____

By ____________________

Name *Date*

DIVISION 1 WORKSHEET 8

Division 1 explains the general conditions of the contract. It is primarily a legal document delegating the responsibilities of the Architect, Owner and Contractor. Under General Conditions, Index to the Articles of the General Conditions, are listed forty-two topics. Study them carefully and review each section and paragraph under Division 1 in your Specification Textbook. Some of these sections and paragraphs may not be needed for your particular project.

On this Worksheet all the sections are listed for you. Draw a line through those sections not applicable to your particular project.

Division 1 General Conditions

Index to the Articles of the General Conditions

1. Definitions
2. Execution, Correlation and Intent of Documents
3. Detail Drawings and Instructions
4. Copies of Drawings Furnished
5. Order of Completion
6. Drawings and Specifications on the Work
7. Ownership of Drawings
8. Contractor's Understanding
9. Materials, Appliances, Employees
10. Royalties and Patents
11. Surveys, Permits and Regulations
12. Protection of Work and Property
13. Inspection of Work
14. Superintendence; Supervision
15. Changes in the Work
16. Extension of Time
17. Claims for Extra Cost
18. Deduction for Uncorrected Work
19. Delays and Extension of Time
20. Correction of Work before Final Payment
21. Suspension of Work
22. The Owner's Right to do Work
23. The Owner's Right to Terminate Contract
24. Contractor's Right to Stop Work or Terminate Contract
25. Removal of Equipment
26. Use of Completed Portions
27. Responsibility for Work
28. Payments Withheld–Prior to Final Acceptance of Work
29. Contractor's Liability Insurance
30. Surety Bonds
31. Damage Claims
32. Liens
33. Assignment
34. Rights of Various Interests
35. Separate Contracts
36. Subcontracts
37. Architect's or Engineer's Status
38. Architect's or Engineer's Decisions
39. Arbitration
40. Lands for Work
41. Cleaning Up
42. Payments

Note: Each division under the Specification Index Division is numbered 1 through 36A and is followed by a description of that particular division. All Divisions must start on a new page. The pages contain the number of the division and its numerical place in the division. For example: Division 1, page 1, (1-1), Division 1, page 2, (1-2), etc.

Name ____________________ Date ____________________

DIVISION 1, SECTIONS 1 THROUGH 42

WORKSHEET 9

Section 1. Definitions

a. The Contract Documents consist of:

  1)
  2)
  3)
    a)
    b)
    c)
  4)
  5)
  6)
  7)
  8)
    a)
    b)
    c)

b. The __________ , the __________________ and the _______________ are those mentioned as such in the Agreement. they are treated, throughout the ____________________ as if each were of the singular number and masculine gender.

c. Wherever in this ____________ the word __________________ is used it shall be understood as referring to the _______ _______________ of the _________________ , acting personally or through an assistant duly authorized in writing for such act by the ____________________ .

d. Written notice shall be deemed to have been duly served if delivered in person to the individual, or to a member of the _________________ or to an officer of the _________________ for whom it is intended, or if delivered at or sent by registered mail to the ______________________ business address known to him who gives the notice.

e. The term _________________ , as employed herein, includes only those having a direct contract with the ____________ and it includes one who furnishes material worked to a special design according to the plans or specifications of this work, but does not include one who merely furnishes material not so worked.

f. The term ______________ of the _________________ or _________________ includes _______________ or _________________ or _________________ , _________________ , _________________ , or other facilities necessary to complete the ___________________ .

g. All time limits stated in the _________________ are of the essence of the _________________ .

h. Working days are defined as those days when the temperature is above the _____________________________ for the work being done, but not including ______________ . _________________ or those days when the _________________ ____ .

i. Wherever it is permitted in these ____________________ to substitute the ______________________ of a designated brand or grade of materials or workmanship, it is understood that in such case the written approval of the ____________ for such substitution must be obtained especially for this work.

Section 2. Execution, Correlation and Intent of Documents

The ____________________ shall be signed in _________________ by the ______________________ and the _______ _______________ . In case the ___________________ and the ___________________ fail to sign the _________________ . ____________________ the _________________ shall identify them.

The ____________________ are complimentary, and what is called for by any one shall be as binding as if called for by all. The intention of the ________________ is to include all ________________ and ______________ , ____________ and ______________ necessary for the proper execution of the work. It is not intended, however, that materials, or work not covered by or properly inferable from any __________ , ______________ , ______________ or ______________ of the specifications shall be supplied unless distinctly so noted on the drawings. ___________ or ______________ described in words which so applied have a wellknown technical or trade meaning shall be held to refer to such recognized standards.

Section 3. Detail Drawings and Instructions

The ________________ shall furnish, with reasonable promptness, additional instructions, by means of ____________ or otherwise, necessary for the proper execution of the work. All such ____________ and ______________ shall be consistent with the ____________________ , true developments thereof, and reasonably inferable therefrom.

Section 4. Copies of Drawings Furnished

Unless otherwise provided in the ________________ , the ______________ shall furnish to the ____________ , free of charge, all ____________________________ reasonably necessary for the execution of the work.

Section 5. Order of Completion

The ________________ shall submit, at such times as may be requested by the ______________ , schedules which shall show the order in which the ______________ purposes to carry on the work with dates at which the ____________ will start the several parts of the work and estimated dates of completion of the several parts.

Section 6. Drawings and Specifications on the Work

The _____________ shall keep one copy of all _____________ and ____________ on the work, in good order, available to the _____________ and his representatives.

Section 7. Ownership of Drawings

All ___________ , ______________ and copies thereof furnished by the ________________ are his property. They are not to be used on other work, and, with the exception of the ________________________ , are to be returned to him on request, at the completion of the work. All models are the property of the _____________________ .

Section 8. Contractor's Understanding

It is understood and agreed that the ________________ has, by careful examination, satisfied himself as to the ______________ and ________________ of the ___________ , the ______________ of the _________________ , the character, ______________ and ______________ of the ______________ to be encountered, the character of ________________ and ____________________ needed preliminary to and during the prosecution of the work, the general and local conditions, and all other matters which can in any way affect the work under this ____________ . No verbal agreement or conversation with any _______ , ___________ or _______________ of the ______________ either before or after the execution of this _________________ , shall affect modify any of the terms or obligations herein contained.

Section 9. Materials, Appliances, Employees

Unless otherwise stipulated, the ______________ shall provide and pay for all ______________, ______________, ______________, ______________, ______________, ______________, ______________, ______________ and other facilities necessary for the execution and completion of the work.

Unless otherwise specified, all materials shall be new and both ______________ and ______________ shall be of a good quality. The ______________ shall, if required, furnish satisfactory evidence as to the ______________ and ______________ of ______________.

The ______________ shall at all times enforce strict discipline and good order among his ______________, and shall not employ on the work any ______________ or any one ______________.

Section 10. Royalties and Patents

The ______________ shall pay all royalties and license fees. He shall defend all ______________ or ______________ for infringement of any patent rights and shall save the ______________ harmless from ______________ on account thereof, except that the ______________ shall be responsible for all such loss when a particular process or the product of a particular manufacturer or manufacturers is specified, but if the ______________ has information that the process or article specified is an infringement of a patent he shall be responsible for such loss unless he promptly gives such information to the ______________.

Section 11. Surveys, Permits and Regulations

The ______________ shall furnish all surveys unless otherwise specified. ______________ and ______________ of a temporary nature necessary for the prosecution of the work shall be secured and paid for by the ______________, ______________, ______________, and ______________ for permanent structures or permanent changes in existing facilities shall be secured and paid for by the ______________, unless otherwise specified.

The ______________ shall give all notices and comply with all laws, ______________, ______________ and ______________ bearing on the conduct of the work as drawn and specified. If the ______________ observes that the drawings and specifications are at variance therewith, he shall promptly notify the ______________ in ______________, and any necessary changes shall be adjusted as provided, in the Contract for ______________. If the ______________ performs any work knowing it to be contrary to such laws, ______________, rule and ______________, and without such notice to the ______________ he shall bear all cost arising therefrom.

Section 12. Protection of Work and Property

The ______________ shall continuously maintain adequate protection of all his work from damage and shall protect the ______________ from injury or loss arising in connection with this ______________. He shall make good any such damage, injury or ______________, except such as may be directly due to errors in the ______________ ______________ or caused by ______________ or ______________ of the ______________. He shall adequately protect adjacent property as provided by ______________ and the ______________ ______________. He shall provide and maintain all passage ways, ______________, ______________ and other facilities for protection required by ______________ or ______________.

In an emergency affecting the safety of life or of the work or of adjoining property, the ______________, without special instructions or authorization from the ______________, is hereby permitted to act, at his discretion, to prevent such threatened loss or injury, and he shall so act, without appeal, if so instructed or authorized. Any compensation, claimed by the ______________ on account of emergency work, shall be determined by ______________ or ______________.

Section 13. Inspection of Work

The ______________ and his representatives shall at all times have access to the work wherever it is in preparation or progress and the ____________ shall provide facilities for such access and for inspection.

If the specifications, the ________________ instructions, laws, ordinances or any public authority require any work to be specially tested or approved, the ______________ shall give the ____________ timely notice of its readiness or inspection, and if the inspection is by another authority than the ____________, of the date fixed for such inspection. Inspections by the ____________ shall be promptly made, and where practicable at the source of supply. If any work should be covered up without approval or consent of the ____________, it must, if required by the ___________, be uncovered for examination at the ____________ expense.

Re-examinations of questioned work may be ordered by the ______________ and if so ordered the work must be uncovered by the ___________. If such work be found in accordance with the ____________________________ the ____________ shall pay the cost for re-examination and replacement. If such work be found not in accordance with the Contract Documents the ________________ shall pay such cost, unless he shall show that the defect in the work was caused by another ___________, and in that event the _____________ shall pay such cost.

Section 14. Superintendence; Supervision

The _________________ shall keep on his work during its progress a competent superintendent and any necessary assistants, all satisfactory to the __________. The _____________ shall not be changed except with the consent of the ___ __________ unless the ___________ proves to be unsatisfactory to the ___________ and ceases to be in his employ. The __________ shall represent the ___________ in his absence and all directions shall be confirmed in writing to the _______ __________. Other directions shall be so confirmed on written request in each case. The ___________ shall give efficient supervision to the work, using his best skill and attention.

If the ____________, in the course of the work, finds any _________ between _________________ and the ____________ of the ___________, or any ____________ or _____________ in _____________ or in the __________________ as given by ______________ and ____________, it shall be his duty to immediately inform the _______________, in _______________, and the _______________ shall promptly verify the same. Any work done after such discovery until authorized, will be done at the _________________ risk.

Neither party shall employ or hire any employee of the other party without ______________________.

Section 15. Changes in the Work

The ___________, without invalidating the _____________, may order extra work or make changes by ________, ___________ or _____________ from the work, the _______________________ being adjusted accordingly. All such work shall be executed under the conditions or the _______________________ except that any claim for extension of time caused thereby shall be adjusted at the time of ________________.

In giving instructions, the ____________ shall have authority to make minor changes in the work, not involving extra _______________, and not inconsistent with the _______________ but otherwise, except in an emergency endangering _________ or __________ shall be made unless in pursuance of a written order by the _____________ and no claim for an addition to the ________________ shall be valid unless so ordered.

The value of any such work or change shall be determined in one or more of the following ways:

a.

b.

c.

If none of the above methods is agreed upon, the __________, provided he receives an order as above, shall proceed with the work. In such case and also under case __________, he shall keep and present in such form as the __________ may direct, a correct account of the __________ certify to the amount, including reasonable allowance for __________ and __________, due to the __________. Pending final determination of value, payments on account of changes shall be made on the __________ .

Section 16. Extension of Time

Extension of time stipulated in the __________ for completion of the work will be made if and as the __________ may deem proper; when work under extra work order as hereinbefore provided is added to the work under this __________ ; then the work is __________ as provided in Section 20; and when the work of the __________ is delayed on account of __________ or __________ of the __________, and which were not the result of his fault or negligence. Extension of time for completion will also be allowed for any delays in the __________ caused by any act or neglect of the __________ or of his employees or by other __________ employed by the __________, or delay due to an __________, or by any delay in the __________ and necessary information by the __________, or by any other cause which in the opinion of the __________ entitles a __________ to an __________. __________ and other labor disputes shall also be cause for an extension of time.

The __________ shall notify the __________ promptly of any occurrence or conditions which in the __________ opinion entitles him to an __________. Such notice shall be in __________ and shall be submitted in ample time to permit __________ and the __________ of the __________. The __________ shall acknowledge receipt of the __________ notice within __________ days of its __________. Failure to provide such notice shall constitute a __________ by the __________ of any such claim.

Section 17. Claims for Extra Cost

If the __________ claims that any instructions by __________ or otherwise involve __________ under this __________, he shall give the __________ written notice thereof within __________ days after the receipt of such instructions and in any event before proceeding to execute work, except in emergency endangering __________ or __________, and the procedure shall then be as provided for changes in the work. No such claim shall be valid unless so made.

Section 18. Deductions for Uncorrected Work

If the __________ deems it inexpedient to correct work injured or done not in accordance with the __________, an equitable deduction form the __________ shall be made therefore.

Section 19. Delays and Extension of Time

If the __________ be delayed at any time in the progress of the work by any act or neglect of the __________, or of his employees, or by an other __________ employed by the __________, or by __________ ordered in the __________, or by __________, __________, __________, __________, __________ or any causes beyond the __________ control, or by __________ by the __________ pending arbitration, or by any cause which the __________ shall decide to justify the delay, then the time of completion shall be extended for such reasonable time as the __________ may decide.

No such extension shall be made for delay occurring more than __________ days before claim therefore is made in ____________ to the ______________. In the case of a ________________ of delay only one claim is necessary.

If no schedule or agreement stating the dates upon which drawings shall be furnished is made then no ____________ shall be allowed on account of failure to ____________________ until two weeks after demand for such drawings and not then unless such claim be reasonable.

This ____________ does not exclude the ___________ or ___________ by either party under other provisions in the ____________________.

Section 20. Correction of Work before Final Payment

The _____________ shall promptly remove from the premises all ________________ by the ___________ as failing to conform to the ______________, whether incorporated in the work or not, and the ____________ shall promptly replace and re-execute his own work in accordance with the ______________ and without expense to the ______ _____________ and shall bear the expense of making good all work of other ____________ destroyed or damaged by such removal or replacement.

If the ___________ does not remove such condemned ____________ and _____________ within a reasonable time, fixed by written notice, the __________ may remove them and may __________ the material at the expense of the _______________. If the ______________ does not pay the _____________ of such removal within _____________ days' time thereafter, the _____________ may, upon _________ days ____________ notice, sell such materials at _____________ or at _______________ and shall account for the net proceeds thereof, after deducting all the ________ and __________ that should have borne by the _________________.

Section 21. Suspension of Work

The ____________ may at any time suspend the work, or any part thereof by giving _________ days' notice to the _____________ in writing. The work shall be resumed by the ___________ within ___________ days after date fixed in the Written notice from the ______________ to the ___________ to do so. The ________ shall reimburse the __________________ expense incurred by the ________________ in connection with the work under this Contract as a result of such suspension.

But if the work or any part thereof shall be stopped by the notice in writing aforesaid, and if the ___________ does not give notice in writing to the _____________ to resume work at a date within _________ days of the date fixed in the written _______ to suspend, then the ____________ may abandon that portion of the work so suspended and he will be entitled to the estimates and payments for all work done on the portions so abandoned if any.

Section 22. The Owner's Right to do Work

If the ____________ shall neglect to prosecute the work properly or fail to perform any provision of this ________, the _____________, after _______ days' written notice to the _________, may, without prejudice to any other remedy he may have, made good such deficiencies and may deduct and delete the cost thereof from the __________ then or thereafter due the ____________________.

Section 23. The Owner's Right to Terminate Contract

If the ______________ should be adjudged a __________, or if he should make a general assignment for the benefit of his ____________, or if a receiver should be appointed on account of his insolvency, or if he should persistently or repeatedly refuse or should fail, except in cases for which extension of time is provided, to supply enough properly skilled

workmen or proper materials, or if he should fail to make prompt payment to subcontractors or for material or__________, or persistently disregard laws, ____________ or the ____________ of the __________, or otherwise be guilty of a substantial violation of any provision of the contract, then the ______________, upon the certificate of the _________ that sufficient cause exists to justify such action, may, without prejudice to any other right or remedy and after giving the ______________ days; written notice, terminate the employment of the __________ and take possession of the ______________ and of all ______________, __________, and ____________ thereon and finish the work by whatever method he may deem expedient. In such a case the _________ shall not be entitled to receive any further payment until the work is finished. If the unpaid balance of the contract price shall exceed the _________ the work, including compensation for additional ____________ and ______________ services, such excess shall be paid to the ______________. If such expense shall exceed such unpaid balance, the ___________ shall pay the difference to the ______________. The expense incurred by the ____________ as herein provided, and the damage incurred through the ______________ default, shall be certified by the ____________________.

Section 24. Contractor's Right to Stop Work or Terminate Contract

If the work should be stopped under an order of any court, or other public authority, for a period of ________months, through no act or fault of the ___________ or of anyone ___________ by him, or if the _________ should fail to issue any ____________ within _______ days after it is due, or if the ____________ should fail to ______________ within ________ days of its maturity and presentation, any sum certified by the ____________ or awarded by ________, then the ______________ may, upon _______ days' written notice to the __________ and the __________, stop work or terminate this ______________ and recover from the ____________ payment for all work executed and any loss sustained upon plant or materials and reasonable profit and damages.

Section 25. Removal of Equipment

In the case of ____________ of this ___________ before completion from any cause whatever, the __________, if notified to do so by the _________, shall promptly remove any part of all of his equipment and supplies from the ______________ of the ___________, failing which the ____________ shall have the right to remove such ______________ and ______________ at the expense of the ______________.

Section 26. Use of Completed Portions

The ________________ shall have the right to take possession of and use any __________ or ______________ portions of the work, notwithstanding the time for completing the entire work or such portions that may not have expired, but such taking possession and use shall not be deemed an acceptance of any work not completed in accordance with the ________________________. If such prior use increases the cost of or ____________ the ______________ shall be entitled to such extra __________ or ____________ or both, as the ______________ may determine.

Section 27. Responsibility for Work

The _____________ assumes full responsibility for ___________ and ___________ used in the construction of the work and agrees to make no claims against the ___________ for damages to such _________ and __________ from any cause except ____________ or ____________ act of the ____________. Until its final acceptance, the _______________ shall be responsible for damage to or destruction of the ___________ (except such work ____________________ as set forth in Section 26.) He shall make good all work damaged or destroyed before acceptance.

Section 28. Payments Withheld–Prior to Final Acceptance of Work

The ______________ may withhold or, on account of subsequently discovered evidence, nullify the whole or a part of any ___________ to such extent as may be necessary to ________________ himself from ________________ on account of:

a.
b.
c.
d.
e.

When the above grounds are removed or the _____________ provides the ____________ satisfactory to the _____ __________, which will protect the ______________ in the amount withheld, payment shall be made from amounts withheld, because of them.

Section 29. Contractor's Liability Insurance

The ________________ shall secure and maintain such insurance policies as will protect himself, his ___________, and unless otherwise specified, the _________, from claims from __________________, death, or ________________ which may arise from operations under this ________________ whether such operations be by himself or by any _______ ___________ or any one employed by them directly or indirectly. The following insurance policies are required:

a.
b.
c.

________________ of such ____________ shall be filed with the ____________ or __________, and shall be subject to approval as adequacy and protection. Said _______________ or _____________ shall contain a ___________ day notice of ______________.

Section 30. Surety Bonds

______________ shall be required to furnish and pay for _________ percent ___________ for entire contract price and also furnish and pay for bond for payment of _________ and ______________.

Section 31. Damage Claims

The _____________ shall defend, _____________ and save harmless the ___________, its _________________, ______________, _____________, and ___________ against and from all ___________, ___________, ___________, ____________, ___________, ___________, ___________ and ___________ of every kind and description, and from all damages to which the ____________ for any of its _______________, _________, _________________ and employees may be subjected by reason of _____________ to the personal property of others resulting from the performance of the project or through any __________ or _____________________, _______________ or _______________ used by the _________________ in the project, or through any act or omission on the part of the ________________ or his ______________, _____________, or _____________; that he shall further defend, indemnify and save harmless

the ______________, its __________, ____________, ______________ and ______________ from all ________________ and ____________ of any kind or __________ whatsoever which may be brought or instituted by any ______________, ________________ or ______________, who has performed work or furnished materials in or about the project or by, or on account of, any ______________ or amount recovered for an ______________ or ______________, ________________ or ________________.

Section 32. Liens

Neither the final payment nor any part of the retained percentage shall become due until the __________, if required, shall deliver to the ____________ a complete release of all ______________ arising out of this __________, or ____________ in full in lieu thereof and, if required in either case, an ____________ that so far as he has knowledge or information the ______________ and ______________ include all the ____________ and ____________ for which a ____________ could be filed; but the ________ may, if any ______________ refuses to furnish a ________________ or ________________ in full, furnish a ______________ satisfactory to the ______________, to indemnify the ________________ against any ________________. If any ______________ remains unsatisfied after all payments are made, the ______________ shall refund to the ______________ all moneys that the latter may be compelled to pay in discharging such a __________, including all cost and a reasonable attorney's-fee.

Section 33. Assignment

Neither party to the ______________ shall assign the ______________ or sublet it as a whole without the written consent of the other, nor shall the ____________ assign any moneys due or to become due to him hereunder without the previous written consent of the ____________________.

Section 34. Rights of Various Interests

Wherever work being done by the ____________________ or by other ______________ is continguous to work covered by this ____________ the respective rights of the various interests involved shall be established by the __________, to secure the completion of the various portions of the work in general harmony.

Section 35. Separate Contracts

The ____________ reserves the right to let other ______________ in connection with this work. The ____________ shall afford other ______________ reasonably opportunity for the ____________ and ______________ of their ________ ____________ and the execution of their work, and shall properly connect and coordinate his work with theirs.

If any part of the ________________ depends for proper execution of results upon the work of any other __________, the ____________ shall inspect and promptly report to the ________________ any defects in such work that render it unsuitable for such proper execution and results. His failure so to inspect and report shall constitute an acceptable of the other ________________ as fit and proper for the reception of his work, except as to defects which may develop in the other ________________ after execution of his work.

To insure the proper execution of his subsequent work the ______________ shall measure work already in place and shall at once report to the ______________ and ____________ between the ______________ and ________________.

Section 36. Subcontracts

The ____________ shall, as soon as practicable after the signature of the ____________ notify the ____________ in writing the names of ____________ proposes for the work and shall not employ any that the ____________ may within a reasonable time object to as ____________ or ____________ .

The ____________ agrees that he is as fully responsible to the ____________ for the acts and omissions of his ____________ and of persons either directly or indirectly employed by them, as he is for the acts and omissions of persons ____________ employed by him.

Nothing contained in the ____________ shall create any contractural relation between any ____________ and the ____________.

Section 37. Architect's or Engineer's status

The ____________ shall have general supervision and direction of the ____________. He has authority to ____________ the work whenever such stoppage may be necessary to insure proper execution of the ____________. He shall also have authority to reject all ____________ and ____________ which do not conform to the ____________, to direct the application of forces to any portion of the work, as in his judgment is required, and to order the force increased or diminished, and to decide questions which arise in the ____________ .

Section 38. Architect's or Engineer's Decisions

The ____________ shall, within a reasonable time after their presentation to him, make decisions in writing on all ____________ or the ____________ or the ____________ and on all other matters relating to the execution and progress of the work or the interpretation of the ____________.

All such decisions of the ____________ shall be final except in cases where time and/or financial consideration are involved, which, if no agreement in regard thereto is reached, shall be subject to ____________ .

Section 39. Arbitration

Demand for ____________. Any and all disputes arising out of, under, or in connection with the ____________ or for a ____________ thereof, shall be submitted to ____________ in accordance with the rules of ____________ , ____________ upon demand of either party to the dispute.

Section 40. Lands for Work

The ____________ shall provide as indicated on drawings and not later than the date when needed by the ____________ the lands upon which the work under this ____________ is to be done, rights-of-way for access to same, and such other lands which are designated on the drawing for the use of the ____________. Any delay in the furnishing of these lands by the ____________ shall be deemed proper cause for an equitable adjustment in both ____________ price and time of completion.

The ____________ shall provide at his own expense, and without ____________ to the ____________ any additional land and access thereto, may be required for temporary ____________ , or for ____________ .

Section 41. Cleaning Up

The ______________ shall, as directed by the ____________, remove at his own expense from the ________________ and from all ______________ and ________________ property all, ____ and __________________ resulting from his operations.

Section 42. Payments

___________________ will be made on monthly estimates rendered by the ______________ and approved by the ___________________. The ______________ shall submit his estimate to the ____________________________ at least ____________ days prior to the _________________ of the month. This estimate shall be broken down in accordance with the itemized complete cost shown in ______________________.

After final completion of the work, the _____________________ shall make a ___________________________ or _______________________ recommending that the work be approved and the _____________________ paid.

Name ______________________________ Date ______________________________

DIVISION 2–SPECIAL CONDITIONS

WORKSHEET 10

Division 2 specifies the conditions relating to but not necessarily a part of the work. These conditions are necessary to facilitate the completion of the project. They are all tied to some extent to various items in your specifications.

Read Division 2, "Special Conditions," in your textbook, then complete this worksheet by filling in only the blank spaces that specify conditions relating to your project.

Division 2–Special Conditions

1. *Time of Completion*

____________________ shall commence work within _________ days of written notice to proceed. The number of calendar days required to complete the work shall be stated in the ________________________.

2. *Examination of Site*

Before bidding the work each contractor shall inform himself fully as to all ___________________________________.

3. *Laying Out Work*

The Contractor shall at his own expense employ an _______________________ to give the required lines, points and levels.

4. *Temporary Heat*

Should the completion of the building proceed into _____________ weather the ____________ will permit the use of the heating plant, or such portion of it as may have been installed, by the General Contractor at his own risk and responsibility. All direct material, ____________, and _________________ at the project in connection with such damage to the heating plant or the building shall be assumed by him.

5. *Temporary Utilities*

a. *Water.* Contractor shall pay for water used for his ___________________________________.

b. *Electrical.* General Contractor must pay for electricity used on the ____________. Electrical Contractor, if required, shall provide temporary connections for the _______________________ – ____________ _________________.

6. *Temporary Toilets*

General Contractor shall provide and maintain a toilet for the use of workmen. When job is complete _______________ and waste shall be disposed of in manner agreeable to ______________________________________.

7. Scope of Work and Work Not Included in Construction Contract

The work to be included in this contract includes all ________ and __________________ necessary for and reasonably incidental to the erection of the new building complete as shown and specified in the ________________________________ and ____________________________ .

8. Storage of Materials

Contractor may store materials on the _________ . Contractor is responsible for all material until building completion and possession by ______________________ .

9. Safety Requirements

Precautions shall be exercised at all times for the protection of ______________ and ________________________ .

10. Samples Required

Samples are required as specified under the various divisions of the work.

Problems of Application

Problem 1. The building is ready for plastering, but the temperature within it is too cold for the contractor to proceed with the work, so he starts the boiler to generate heat for the building. After the building is completed, the general contractor refuses to pay the fuel bill incurred in supplying this heat, saying that because he heated the building, he had been able to complete the work several weeks ahead of schedule, which saved the owner a considerable amount of money.

Who is responsible for this bill? Why?

Problem 2. The architect and engineer both agreed that the temporary toilet facilities furnished by the general contractor (he had simply dug a trench and placed an outhouse over it) were O.K. for use by the workmen. However, these facilities were not in conformity with the City Code and the general contractor was forced to install a water closet and connect it to the city water and sewer system. The general contractor maintained that he should not have to pay this extra expense since the architect and engineer both O.K'd his original facilities.

Who is liable for this extra expense and why?

Name Date

DIVISION 3–ALTERNATES

WORKSHEET 11

Division 3 contains all the alternates that are included in the contract documents and shown on the plans. In your Specification Textbook, G-1 through G-5 and E-1 are included as alternates. In many cases, perhaps, only one or two alternates are used. In the general contract, for instance, you could have two or three alternates and none in the electrical or plumbing contracts. Or, you might not have any in the general contract and have two in the electrical and one in the plumbing contract. This division, then, must always be tailored to meet your project requirements.

Division 3–Alternates

Problems of Application

Alternate 1. You are specifying "X" type windows and the owner thinks that "Y" type window would cost less and be more serviceable. Prepare an alternate showing the merits of "Y" type windows so that the owner can reach a decision based on facts.

Alternate 2. The owner also feels that "Y" insulation could be used to better advantage in his building than the one you are specifying. Prepare an alternate for "Y" insulation so that he can compare the values of "X" and "Y" insulation.

Name *Date*

DIVISION 4–DEMOLITION

WORKSHEET 12

Sometimes it is necessary to tear down or remove an old building before a new one can be constructed on the site. The owner usually decides whether the general contractor or a private party will dispose of the building and its furnishings. If it is to be the contractor's responsibility to demolish the building and dispose of the furnishings and refuse, then the conditions pertinent to the demolition and disposal become a part of the General Contract.

Problem of Application

Problem. A ten story, fifty-year-old hotel had to be removed before a new building could be constructed on the site. The owner decided that it would be the contractor's responsibility to dispose of the furnishings, building and refuse. Included in the furnishings were such items as beds, chests, curtains, kitchen equipment, lavatories, bathtubs, toilets, etc. Any money realized from the sale of the furnishings or any part of the building was to be kept by the contractor.

If you were the contractor, would you sell any of the furnishings? If so, how would you handle their sale? Do you think the bricks, steel, etc., used in the construction of the old building would have any resale value? Do you think the manner in which the contractor planned to dispose of the furnishings, building and refuse would have any bearing on the base bid of his contract? Why?

Name ________________________________ Date ________________________________

DIVISION 5–EXCAVATING–GRADING   WORKSHEET 13

The paragraphs under excavating and grading in your Specification Text pretty well outline the contractor's responsibilities in this area. After reading these paragraphs, complete this worksheet according to your project requirements.

Division 5–Excavating–Grading

1. General Notes

a. The ______________ shall perform all ______________, ______________, ______________, and rough and finish grading as indicated on the drawings and as hereinafter described. The __________ shall, as a part of his contract, perform all trenching for footings all _________ against walls, and all fill under ______ floors, _________, or ______________ resting directly on ground.

b. Excavating, backfilling for ______, ______, ______, ____________, ____________, and ___________ work are to be included under their respective divisions.

c. Remove and dispose off-site unsuitable and excess ______________.

d. Seeding and sodding not included in this contract.

2. Subsurface Soil Data (See also Special Conditions)

If any serious obstacles or conditions are encountered the __________ shall be notified immediately and adjustments or ___________ worked out.

3. Excavating and Filling

a. Excavate to the dimensions shown on the various drawings, and do all other excavation necessary to fully carry out the work shown or herein specified. Provide ________ wherever necessary. All footings ______________ are to be excavated to the exact depths and widths required, with the sides ____________ and the bottom ___________. In no case shall bottom of footing be less than _____________ below natural grade line and _____________ below finish grade line.

b. The ___________ shall make such excavation and provide such drainage ________ as will at all time keep the building site free from standing or running water, and will be held fully responsible for the consequences or failure to do so.

c. Filled areas under ____________ are to be "filled" with ____________ and ____________________.

d. Where trenches for footings are not straight and __________________ will be required. Attention must be given to the top edge of the ___________ to prevent cave-ins.

e. Gravel-sand fill shall be placed in ____________ layers and shall be uniformly moist to permit _____________ with ________ __________________________ and __________________________. Gravel-sand sample for use in "filled" areas shall be approved by the ______________.

f. If suitable bearing for foundations are not encountered at depth indicated ______________________ to immediately notify ______________ and not proceed further until instructions given.

g. Care must be taken to prevent damage to existing sewage drainage tile and water supply __________ on existing ___ ________. Work on the existing system is to be accomplished by the _______ ___________________.

See site plan for approximate location of piping and _________ _______________. Necessary measurements made to establish addition of volume of excavation.

h. Protect ________ of excavation from ________. No foundations or slabs on frozen ground.

i. Fill excess ____________ under ________________ with __________________________.

4. *Grading*

a. Grading of the site in general, shall be done by the __________________. Soil taken from the building excavation and from the ______________ operations shall be used in grading as herein specified, to the grades and limits indicated on the plot plan. Provide additional ____________ for ____________________ if necessary.

b. Rough grading shall be completed as soon as practicable but final or finish _______________ shall be delayed until construction is nearing _________________________.

c. Remove debris from excavations before _____________; ____________________________________ must be free from _________, ___________, and other debris. Deposit in layers not exceeding ______________ under slabs, pavements or other surface; __________ other areas, compact each layer fill uniformly on both sides of foundation walls.

Problems of Application

Problem 1. While excavating at the building site, the contractor came upon a six-inch iron pipe of undetermined length, buried about five feet into the ground. If you were the contractor and this pipe did not interfere with the footings or the building in any way, what would you do with the pipe, if anything?

Problem 2. Seeding and sodding are not included in the General Specification, Division 5. However, now that the building is almost completed, the owner has changed his mind and decided to have the contractor sod the area surrounding the building. What steps must be taken to protect the owner, architect and contractor before the contractor complies with the owner's request?

Name *Date*

DIVISION 6–PILES AND SPECIAL FOUNDATIONS

WORKSHEET 14

Sometimes the load-bearing capacity of the earth upon which the building is to rest is not sufficient to hold the weight of the building. Therefore, piles and special foundations are required to hold it in place. Piles can be made of steel, wood, or concrete and are pounded or set into the ground until a firm base is reached. This division outlines the contractor's responsibility in having the earth tested to insure a solid foundation for the building. This testing is usually done by a civil or structural engineer at the owner's expense and is not considered as a part of the regular expense. However, it is a related cost and, therefore appears in the specifications.

Division 6–Piles and Special Foundations

Problems of Application

Problem. The congregation decided to build a church on land that had been donated to them. The plans were completed; the contract awarded; and the contractor proceeded to have the necessary analysis and compression tests made of the earth as specified in his contract.

The civil engineer hired to test the soil reported that the earth was spongy where one corner of the church would rest and, therefore, requested that steel piles be driven into the earth at that point to a depth of 35 feet.

The contractor, however, did not agree with his findings and on his own, ordered a new analysis and compression test made of the earth.

According to the contract, Division 6, of your Specification Textbook, did the contractor have the authority to order new tests? Who would be responsible for the costs of the additional tests and why?

Name Date

DIVISION 7–CONCRETE WORK

WORKSHEET 15

Concrete work is included in every building in some form or another, for example; floors, footings, foundations, etc. It is important that you read and thoroughly understand every paragraph under "Concrete Work," Division 7, in your Specification Textbook. You will note that for each use included, detailed descriptions of the materials used are given, because the strength and durability of concrete depend directly upon the materials used and their mixture.

The expansion and contraction of concrete necessitates the use of expansion and scored joints. Their use must be outlined in your specification.

Weather conditions must be taken into consideration since they affect the finished product; therefore, drying time and the outside temperature must be included in your specification.

Division 7–Concrete Work

Problem of Application

Problem. A contractor used sand and gravel extensively in preparing the bedding for the concrete work on your building. Several weeks after he had completed the concrete work and had the proper tests made, it was noticed that the floors in some of the rooms had settled 1/8th of an inch lower than the floors in the hall. A week later the floors were still settling so the architect stopped all work on the building until the reason for the settling could be established.

What do you think caused the settling? Why?

Do you think the work performed under another division of the contract could have a bearing on the work performed in this division? If so, why?

Name Date

DIVISION 8–REINFORCING STEEL WORKSHEET 16

Reinforcing steel is used to prevent temperature and tension cracking. Steel is used because it has the high tensile strength necessary to balance the weak tensile strength of concrete. The two important types of reinforcement are rods, used for example, in beams and columns, and mesh which is used in floors and walks; in roads, rods are often used, as well as mesh, to carry the tension between sections. The correct placing of reinforcing is important and is based upon the principle established by stress analysis that the greatest tensile stress occurs close to thé outside surface.

Division 8–Reinforcing Steel

Problem of Application

Problem. In your Specification Textbook, Division 8, Reinforcing Steel, requires shop drawings to be submitted. What is a shop drawing?
Why are they important to this division?
Who approves them?

Name Date

DIVISION 9–MASONRY

WORKSHEET 17

Considered in this division are brick, concrete tile, solid concrete block, etc. A wide variety of these materials and methods of utilization can be used; therefore, it is extremely important that you include in your specification an accurate description of the material and mortar needed for your project.

Proper weather conditions and curing time must be specified because masonry, like concrete, is affected by the weather.

Masonry is usually cut to allow for fittings to be installed. The proper method should be defined.

It is often necessary to reinforce masonry walls. If used, the type of reinforcing used and its exact placement must be included.

Many times the hollow cores of lightweight masonry are filled with insulation. This too should be included in the specification, if used.

Division 9–Masonry

Problems of Application

Problem 1. Assume that loose filled insulation was poured into the voids of the concrete block and that after the walls were completely laid up, the contractor found it necessary to make holes in the concrete block to facilitate the installation of a light switch.

What effect would this have on the insulation qualities of the wall?

Problem 2. Your Specification Textbook calls for a foam insulation to be installed in the 1" space between the brick and concrete block. However, the owner has now decided that "X" brand has better insulation qualities and is less expensive. The "R" quality of both brands is the same. What other factors must be considered before you can actually decide which insulation is the better of the two?

Name ______ *Date* ______

DIVISION 10–STONE

WORKSHEET 18

Stone is not used for every building; therefore, this Division need not be included in every specification. It is important that you become familiar with it, though, so read this chapter carefully in your Textbook.

Division 10–Stone

Problem of Application

Problem. Assume that the contractor did not order enough stone to do the job or that through careless handling he broke several pieces and had to order more stone to complete the building.

What effect would his inefficiency have on the finished building? Why?

Name Date

DIVISION 11–WATERPROOFING AND DAMP PROOFING

WORKSHEET 19

Buildings constructed on porous ground are subject to seepage; therefore, provisions must be made to protect the building from it. Actually, almost every building constructed today is protected against seepage rather than risking any damage from this source.

One of the methods commonly used for the prevention of seepage damage can be found in your Specification Textbook.

Division 11–Damp Proofing

Problem of Application

Problem. Name two other materials that can be used for damp proofing. Could any of these materials be used in the construction of swimming pools? why?

Name Date

DIVISION 12–STRUCTURAL STEEL

WORKSHEET 20

In this division the responsibilities of the contractor in providing the structure or framework of the building is outlined. There are many systems that can be used; for example, wood, steel, concrete, and pre-stressed concrete. In the Textbook, we have used steel.

Since your plans show only the dimensions, weight, location and shape of all members, any detail of assembly, etc., is usually supplied by the fabricator; therefore, your specifications must outline his responsibility in this area.

All structural material must meet certain rigid specifications as set up by the government or some other regulating body; for example, in your Textbook you will see that we have said that the Structural Steel shall conform to the Standard Specifications for Structural Steel for Buildings as adopted by the American Society for Testing Materials.

Some of the parts to be included in this division are; anchors, bases, beams, purlins, girts, bearing plates, bracing brackets, door frames, expansion joints, floor plates, hangers, lintels, tees, clips, rivets, and bolts.

Use your textbook as a guide in writing these specifications. The Manual of Steel Construction and Sweets Files contains other engineering data that will be helpful.

Division 12–Structural Steel

Problems of Application

Problem. This is probably the most important division in your specification. Why?

Why do you think shop drawings are important to this division?

SUGGESTED READING: Code of Standard Practices for Steel Buildings.

Name *Date*

DIVISION 13–STEEL JOISTS

WORKSHEET 21

This division outlines the contractor's responsibility for furnishing and constructing the supports or joists for the floor or roof system. These supports can be made of steel, wood, concrete, or pre-stressed concrete.

In our Textbook we have specified steel joists and the specification is written expressly for them. You will note too in the Textbook the reference made to Division 8 of the specifications. If this reference had not been made, these conditions, etc., would have been contained in this division. There is also a reference made to Division 1, Paragraph 13. In this manner you can outline the subcontractor's responsibilities in a specific situation without describing it all over again.

Division 13–Steel Joists

Name Date

DIVISION 14–SPECIAL FLOOR-CARPETING WORKSHEET 22

This division allows you to specify materials for special floors that are not used in other parts of the building, such as carpeting. Other materials can be used.

In our Textbook, we have specified carpeting; therefore, the specification is written for carpeting. If you are using another floor covering, you may use the specification as a guide in outlining the contractor's responsibility in furnishing and installing your choice of floor covering.

Division 14–Special Floors

Name Date

DIVISION 15–SPECIAL FLOOR-ROOF DECK

WORKSHEET 23

Today, roof decks can be made of steel, petrical, gypsum, wood, or plastic. Tomorrow, there could be many more materials due to the scientific advances in this field. In our Specification Textbook, gypsum roof decking was specified.

While the material the roof deck is to be constructed of is important and must be carefully considered, the installation is of the utmost importance since it is a safety factor for the occupants of the building. Because the materials and workmanship are so important, shop drawings are required.

This section, then, will cover all materials and equipment necessary for the installation of the roof deck, shop drawings, materials and workmanship.

Division 15–Roof Deck

Name Date

DIVISION 16–TRANSLUCENT STRUCTURAL PANEL SPECIFICATIONS

WORKSHEET 24

Special walls are used today in the construction of many modern buildings. These walls are usually called "Curtain Walls." There are many kinds and brands of curtain walls that can be used. A study of the Curtain Wall Section in the Sweets Catalog will give you an idea of the scope of these walls. It will also provide you with the engineering data needed to write the specifications for your curtain wall.

You will note that in the Specification Textbook we have used Translucent Structural Panels and our specification is written around them. If you are using curtain walls, be sure to write your specification around the kind that you are using.

Division 16–Special Walls

Problem of Application

Problem. Would you use the same heat-gain factors for air-conditioning and heat-loss factors for heating when using curtain walls that you use to heat and air-condition a building constructed with another kind of wall? Why?

Name Date

DIVISION 17–ORNAMENTAL IRON

WORKSHEET 25

This division authorizes the contractor's responsibility in furnishing the iron work required for the project. Ornamental iron is used for stairways and for protective devices; therefore, safety is usually considered first and beauty second. Care must be taken when specifying ornamental iron to make sure that the bars are spaced correctly and all details clearly outlined.

Division 17–Ornamental Iron

Name Date

DIVISION 18–MISCELLANEOUS METAL AND METAL SPECIALTIES

WORKSHEET 26

Miscellaneous metal and metal specialties cover a great many small articles such as foot scrapers, toilet paper holders, and mirrors. Since there are so many kinds to choose from, these items should be detailed in your specification. Using your Specification Textbook as a guide will help you select the items belonging in this division.

Division 18–Miscellaneous Metal and Metal Specialties

Problem of Application

Problem. Why is it important to have a special division for these miscellaneous items?

Name *Date*

DIVISION 19–METAL FRAMES

WORKSHEET 27

The metal doors, windows and frames to be supplied by the contractor are shown in this division. Their construction is also detailed. The necessary hardware, painting and glazing of the metal doors and windows must be included. The doors and windows listed in your Specification Textbook are examples of materials that can be used. Always make sure when writing a specification that the description matches the products that will be used.

Division 19–Metal Frames

Name ______________________ *Date* ______________________

DIVISION 20–SHEET METAL

WORKSHEET 28

Roof flashings are used on all buildings in places where leaks are likely to occur such as around chimneys, dormers, skylights and valleys. Because these flashings are made of sheet metal they are included in this division. The plumber's outlet is an exception. All flashings should be described in your specification as to material and location.

A provision for all sheet metal items required for expansion joints, as detailed on the plans, must be included, also, any galvanized metal gravel stops that are used around the perimeter of some flat roof buildings. Metal hoods for ranges and dishwashers and their description is also a part of this specification.

Division 20–Sheet Metal

Problems of Application

Why are the flashings for the plumber's outlets not included in this division?

Why are measurements taken at the building before the sheet metal is fabricated?

Why aren't these measurements listed in the specification?

Name Date

DIVISION 21–ROOFING

WORKSHEET 29

This is another important division in your specification. A poorly installed roof, or one constructed of inferior materials, can cause severe damage to the interior and exterior of a building and its contents. For this reason many architects will specify only bonded roofs.

Bonded roofs must be carefully constructed since they are insured, in some cases, for a period from two to twenty years against leakage and water damage. During this time, the contractor is liable for any damage that may occur to the building because of inferior roofing materials and poor construction.

There are many kinds of roofs and they should be carefully detailed in the specification. Remember that weather is a factor that must be considered when selecting and applying roofing materials. The Roofing Division in the Sweets Catalog has listed many roofing materials and the engineering data that will be helpful in completing a roofing specification.

Division 21–Roofing

Problems of Application

Problem. In the Specification Textbook, you will see in Division 21, under the heading "Guarantee" and in paragraph b, that a roofer's surety bond was not required. What is a roofer's surety bond and why was it not required in this specification?

Name Date

DIVISION 22–CAULKING AND WEATHERSTRIPPING

WORKSHEET 30

Caulking and weatherstripping are necessary where a part of the building might show penetration of moisture, such as around doors, windows and other similar openings. The contractor's responsibility is outlined in the general notes and the manner of application, the materials and locations of doors, etc., to be stripped or caulked are found in this division.

Division 22–Caulking and Weatherstripping

Name Date

DIVISION 23—CARPENTRY

WORKSHEET 31

All the materials and labor for general and rough carpentry are included in this division; the millwork or finish work is specified in another division.

You would specify here the kind and grades of lumber to be used; the rough hardware such as nails, screws, bolts, insulation, supports, and shelving. These materials vary from job to job.

Your Specification Textbook will serve as a guide in helping you to determine what to include in this division.

Division 23—Carpentry

Name Date

DIVISION 24–MILLWORK

WORKSHEET 32

Finish work for the interior and exterior of the building such as woodtrim, cabinets, wooden doors and windows are included in this division. The materials and the manner in which they are to be installed are all specified. Details such as the installation and storage of materials are included.

Division 24–Millwork

Problems of Application

Problem. If you will compare Divisions 23 and 24, you will note that shelving is included in both divisions. Why?

You will also note that there is no mention made of hardware for the doors and windows in this division. Why? Where is it mentioned?

Name Date

DIVISION 25–METAL LATH, PLASTER DRY WALL AND GYPSUM

WORKSHEET 33

Metal lath is rapidly replacing wooden lath in construction and is mostly specified. There are various types of metal lath that can be used. The kind used and the manner in which it is to be installed are specified here.

Much of the beauty of a building depends upon its finish; therefore, many specifications state that the plastering shall be done only by those having so many years of experience. Designate here (you may have up to three coats of plaster; rough, brown and finish) which coats are to be applied, their composition and the manner in which they are to be applied. The quality of the finish plaster is affected by weather conditions; therefore, the temperature control for its application and curing are included in the specification.

Dry wall gypsum board is also used for the finishing of interior walls. If used, its description, use and application would be detailed in this division.

Division 25–Metal Lath, Plaster, Dry Wall and Gypsum

Name Date

DIVISION 26–SPECIAL COATINGS

WORKSHEET 34

In many buildings special coatings, such as "Polyester Coatings" are used in toilet rooms, hallways, and on walls.

This division designates the areas to be covered, details and gives a description of the material to be used, the preparation of the surfaces and application. For more information on special coatings read Divisions 26 and 26A in your Specification Textbook.

Division 26–Special Coatings

Name Date

DIVISION 27–QUARRY TILE WORKSHEET 35

Quarry tile is not used in every project. However, when it is used, the material to be used and the manner of installation are described here.

Division 27–Quarry Tile

Name *Date*

DIVISION 28—COMPOSITION FLOORING AND COUNTER COVERING

WORKSHEET 36

All labor, materials, equipment and services required for the installation of composition floors and counter tops are included in this division. You will note in your Specification Text that this is another division where the contractor must guarantee his work.

Division 28—Composition Flooring and Counter Covering

Name ________________ *Date* ________________

DIVISION 29–GLASS AND GLAZING

WORKSHEET 37

The glass and glazing for all windows and doors not furnished complete with glass are included in this division. Since much of this work is done by a subcontractor, it is subject to "General Conditions," Sections 35 and 36 in Division 1.

Division 29–Glass and Glazing

Name Date

DIVISION 30–ACOUSTICAL CEILINGS WORKSHEET 38

Acoustical ceilings affect the hearing of sounds throughout a room or hall. There are various ways of installing acoustical ceilings and many of them are installed by subcontractors. The "Suspended Grid System" used in your Specification Textbook is an example of one system. In this division you should describe the areas where it is to be used, designate the materials and outline the application and procedure of the work.

Division 30–Acoustical Ceilings

Name *Date*

DIVISION 31–HARDWARE WORKSHEET 39

A cash allowance is usually allotted for the purchase of finish hardware. The allowance and selection of the items included or excluded are outlined in this division. Division 31, Hardware, in your Specification Textbook, will help you to determine what to include in this division of your specification.

Division 31–Hardware

Name ______________________________ *Date* ______________________________

DIVISION 32–PAINTING AND FINISHING

WORKSHEET 40

This division includes all labor and material necessary for the painting and decorating of the interior and exterior of the building. It also includes the necessary preparation of all surfaces.

Division 32–Painting and Finishing

Problem of Application

Problem. In this division, as in Divisions 26, 26a, 27, 28, and 30, the finish or final touch is specified for each item such as doors, windows, etc. Yet you will note that room numbers are not designated in the specifications for any of these divisions. How, then, does the contractor know what finish to apply where?

Name Date

DIVISION 33–MISCELLANEOUS AND SPECIAL EQUIPMENT

WORKSHEET 41

This division is a catch-all for materials, installations, and equipment that have not been covered in some other division in this specification.

You will note in Division 33 in the Specification Textbook, that we have listed chalkboards, flagpoles, and alternates on basketball equipment. If you were building a church, you could include a cross, bells, etc., in your specification.

Division 33–Miscellaneous and Special Equipment

Name *Date*

DIVISION 34–OPEN

WORKSHEET 42

This particular division is to be used for any special mechanical equipment needed that is not mentioned in any other part of the specification and that must be purchased by the contractor for usage by the owner. A prefabricated walk-in cooler or a drive-in depository window for a bank are examples of this special mechanical equipment. Each of these items is manufactured by a specialty manufacturer and is usually installed by the manufacturer in cooperation with the contractor's craftsmen.

Division 34–Open

Name ______________________________ *Date* ______________________________

DIVISION 35–PLUMBING WORKSHEET 43

Generally the plumbing contract is let out under a separate contract to a Plumbing Contractor. Therefore, in this division, we again list the General Notes, etc. So that you can become familiar with all phases of the plumbing contract, you are required to fill in the spaces left blank. List your materials and outline your installations in the required places. Be sure that you are listing only the materials and installations called for in your building.

Division 35–Plumbing, in your specification textbook, will help you to complete this worksheet. Study it carefully.

Plumbing

1. General Notes

a. The bidder shall carefully examine the ____________ and ________________ , and ____________ the site of construction. The ________ shall fully inform themselves of all the existing conditions and limitations, and shall include in their bid a sum to cover the cost of ____
b. The __________ shall be familiar with the city, county, state and national codes, and to which code takes precedence, and where the drawings and ________________ do not cover such items the ____________________ shall be ________________ for them.
c. It is pointed out that the general conditions and special conditions of the ______________ are considered a part of these specifications and this ________________ .
d. Should bidder find discrepancies in, or omissions from the ____________ and ______________, or should he be in doubt as to their meaning, he should at once notify the ______________ _______ ________________, who will send written instructions to all bidders. Neither the Owner or the Architect or Engineer will be responsible for any ______ instructions.
e. Provide all items, articles, equipment, material, operations, and tools for the erections of the mechanical systems shown on drawings or specified herein, including labor supervision, and incidentals required and necessary to complete the systems for successful operation. Install systems in complete accordance with standards of local, state or national codes governing such work. Place complete systems in a safe and satisfactory condition; adjust all automatic control devices in ____________.
f. All equipment installed in strict accordance with ______________________________ and these ______________ unless otherwise approved in writing. Vertical and horizontal dimensions of all fixtures, units and equipment __________ for installation in spaces provided therefore. _________ ______________ provided for maintenance and repair. Instruct _________ in operation and maintenance; ___________ one set to the _________ of repair parts list and one set instructions on ____________ and ______________ of all equipment installed. All material new and unused.

2. Fees and Permits

a. Pay all ________, ____________, __________ for inspection ____________ ______________ on connection, etc., in connection with the _________. Upon completion of work, furnish _________ with certificate of final inspection from ___________ ___________ having jurisdiction.

3. Drawings

a. Drawings forming a part of these specifications are numbered and described as follows:

Sheet ____________________

4. Data and Measurements

a. Data herein and on drawings is __________ as could be obtained. Absolute accuracy not guaranteed. Obtain __________ __________, __________, __________, etc., at site and satisfactorily adapt work to actual building conditions; installing system generally as shown on plans.

5. Delivery and Storage of Material

a. Provide for __________ and __________ __________ of materials and arrange with other Contractor for introduction into building of equipment too large to pass through __________ __________. Deliver materials at such stages of work as will __________ work as a whole and make and store in such a way as to be easily checked and inspected.

6. Materials and Equipment

a. When make and type definitely specified, __________ must be based on that make and type. When __________ one of two or more types, equally acceptable, __________ of make and type mentioned, but __________ make and type throughout. When specified a certain make or type, "or approved alternate," make and type specifically required __________ written approval of "alternate material" is obtained from Architect or Engineer, prior to delivery on job.

7. Approval Data

a. Approval granted on __________ __________ and __________ __________ __________ rendered as a service only and not considered guarantee of quantities, measurements or building __________; nor construed as __________ the contractor of basic responsibilities under contract.
b. For items __________ from those specified herein and requiring approval, the __________ shall submit complete data within 30 days after contract award. Failure to submit __________ __________ data within the 30 days after the contract award shall be interpreted to mean that the contractor will furnish equipment as specified. Requests for __________ of __________ __________ after the 30 days period will not be considered unless it can be shown that, due to circumstances beyond the Contractor's control the specified material is not available, and cannot be obtained in time to prevent unnecessary delay of construction.
c. For __________ of approval the Contractor shall submit six (6) sets of complete schedules material and equipment proposed for installation. Include catalogs, diagrams, drawings, data sheets and all descriptive literature and data required to enable the Architect or Engineer to determine compliance with construction and contract requirements. Schedules and all data __________ __________ at one time; no __________ will be given to partial or incomplete schedules.
d. Include with data, __________ __________ describing all differences between substitution and item specified; __________ changes required for piping, ducts, wiring space, structure, and all portions of project __________ by substitution.
e. The Contractor will be limited to one approval on a singular piece of equipment. __________ will not be given where the Contractor __________ approval data on one piece of equipment, by more than one manufacturer.

8. Installation of Work

a. Examine __________ and __________ for other work, and if any __________ occur between Plans for this work and Plans for work of others report such discrepancies to Architect or Engineer and obtain written instructions for changes in work of others. Any __________ in work, made necessary through __________ or __________ of Contractor to report such discrepancies, made by and at expense of Contractor. Confer and cooperate with others on work and arrange this work in proper relation with theirs. __________ will be held solely __________ for proper size and location of anchors, inserts, sleeves, chases, recesses, openings, etc., required for proper installation of work. Arrange with proper contractors for building in anchors, etc., do all cutting and patching made necessary by failures or neglect to make

such arrangements with others. Any ________ or ________ subject to directions of Architect or Engineer and not ________ until approval obtained. Damage due to ________ shall be repaired by the ________ . Notify General Contractor of ____________ and location of all openings required for recessed units and equipment items.

9. *Electrical*

a. Furnish all ________ , __________ or ________ __________ , ________ , __________ protective and signalling devices required for operation of equipment specified herein; furnish complete ________ ____________ and ____________ to Electrical Contractor for installation. Wiring diagrams and instructions shall be in ________ of Electrical Contractor within two weeks after contract award. Safety disconnect switches where required shall be provided by the Electrical Contractor.

b. Manual Control shall be ______________________________ starters of proper horsepower and voltage rating. Motors of any rating may be directly ____________ by automatic control devices such as __________ , ________ , ________ ____________ or pressure switches, where such ________ have adequate horsepower and voltage, rating, and proper protection for overloads of motors are incorporated with in the control device. Otherwise the __________ __________ shall be __________ for the operation of the starting and holding coil of a ________________ starter.

10. *Temporary Water Service and Toilet Facilities*

a. Temporary water service and toilet facilities are the responsibility of the ________ ____________, and all facilities shall be at the expense of the General Contractor. However, the General Contractor may request the __________ of the Mechanical Contractor for installation of the __________ ____________ for use during the period of construction, and also for the removal of the equipment when the construction is completed. As soon as construction permits and upon direction of the Architect or Engineer install either permanent or temporary toilet facilities connected to the plumbing systems, with the building, for use during construction.

11. *Trenching and Backfilling*

a. Trenches __________ true to line and grade, ________ properly sheeted, shored, braced where required. Provide accurately contoured bottom for uniform ________ on undisturbed soil for at least ________ pipe circumference for each section of pipe along its entire length, except bell holes excavated progressively with pipe laying. Care should be taken in excavating that __________ or __________ are not injured in any way. Control grading and stacking to prevent ________ ______ flowing into trenches. Any water accumulating shall be removed continually. __________ shall not be done until pipe joints are thoroughly set. After testing lines and inspection, backfill trenches with excavated material free of debris, large clods or stones, 4" layers, uncompacted thickness, moistened and thoroughly tamped, until pipe has covered at least one foot. Remainder excavated material, in 6" layers, moistened and tamped to density undisturbed earth. Flooding trench shall not be __________ . Should subsequent __________ occur, ____________ shall be opened to depth required, refilled, and compact. Pavement or walk cuts shall be restored to original condition.

12. *Piping Installation*

a. *General*–Cut to measured fit at building, installed parallel to walls and ceiling, properly clear all openings, provide required clearance as for operation of doors, windows, access panels, valves, etc., excessive cutting not ____________ . Make changes in direction with fittings, except ______ ________ in soft temper tubing. Interior pipe shall be kept ________ of __________ , ________ ______ and ______ material of any nature. Open ends shall be capped or plugged when work is not in progress. Install to permit ______ ____________ without causing damage to joints, pipe hangers,

units to which connected, structure, or undue noise. Exposed piping shall be run closely as possible to ______________ ______________, piping consealed in wall chases or spaces or furring provided wherever possible. Piping shall be tight and tested before inclosed. Unions shall be made where required for ______________ to facilitate quick repair without disconnection of long lengths or equipment. Valves, trap cleanout, or other part normally ____________ for operation or maintenance not installed in inaccessible place.

b. Drainage and Vent Piping

1. General: Minimum grade horizontal drainage piping shall be ______________________________ ______________________________.

   Change size drain lines with reducing fittings or recessed reducers; change direction with ______________________________ ______________________________.

   except sanitary tees may be used on vertical stacks and short quarter bends or elbows may be used in drainage lines where flow from horizontal to vertical.

2. Underground: Pipes laid hubs facing up-grade. Not less than two lengths of pipe in position, joints finished, earth fill tamped along side pipe, ahead of ________ __________ before poured, except at __________ .

3. Above Ground: All main vertical soil and waste lines shall be ________ full size to and above roof as vents. Where practicable, two or more vent pipes connected together, extended as one pipe through roof, _____ _____ sheet lead extending on roof _____ from pipe in all directions, carried up, over, into top of pipe ______________________ ______________________________ in accordance with manufacturer's recommendations. Vents extended above roof _____ increased in size _____ below under side of roof as required. Minimum size through roof _______. Vent pipes in roof spaces and under floors shall run close as possible to construction, horizontal runs pitched down to stacks without forming ______. Connection of end or circuit vent pipe to vent serving other fixture, at least four feet above flood level fixture served. Branch waste connections to fixture same size fixture outlet unless otherwise shown.

13. Joints and Connections

a. General: All ____________ shall be _______ and _______ tight. Caulking threaded joints or holes will not be permitted. Paint, varnish, putty, etc., will not be permitted on joint until joint is __________ tight. Crosses used on vent pipes, or as specified. Joints in steel pipe shall be made with clean cut ___________, graphite and oil compound applied to make thread only, pipe reamed full size after cutting. Joints in hard copper tubing hard soldered surfaces properly __________ and __________, butt ends reamed, fluxed as recommended by maker of fittings. Flared joints shall be ________ with tubing expanded without splits, using proper flaring tool. Concealed joints in copper water piping, sweat, hard soldered, or brazed. Type of solder for fittings shall be in accordance with data in the National Bureau of Standards Publication," ___ ______________________________ as tabulated by the Copper and Brass Research Association.

b. For prevention of __________ ____________ at connections between pipe of dissimilar metals, such as steel to copper, provide dielectric pipe unions of flange unions similar to ______________________________.

14. Drainage Piping

a. Pipe or fittings with double hubs on same run or double tee branches will _________ be permitted on soil or waste piping. Drilling, tapping or welding soil, drain or vent pipes or use of saddle hubs or bands _________ _________ _________ ____________. Dead ends shall be avoided except where necessary to extend cleanouts to accessible elevations. Slip joints in drainage piping shall be used only in __________ ___________ between trap seal and fixture. Metal contact unions used in drainage piping only in _________ _________ and on inlet side trap.

b. Tile pipe shall have two rounds well tamped rope oakum, fill balance with cement mortar rammed in position, wiped flush with hub. Interior joints shall be ____________ clean with swab large enough to fill pipe, drawn past each joint as made.

Mortar, ____

________________________ . Joints cast iron to tile pipe encased with heavy mass ____________

____ . Clay pipe joint compound CPI-2 used in lieu of mortar for joints, option of Contractor. If used, compound shall be heated and poured in strict accordance with manufacturer's recommendations including application of coal tar film-forming primer whenever wet, oily or cold conditions prevail.

c. Cast iron pipe, picked oakum and lead, 12 ounces lead each inch diameter in each joint made, poured full at one pouring and ________ ______ . Joints lead to cast iron pipe, with extra heavy brass ferrules, extra long, wiped to lead pipe, caulked into hub cast iron coupling ________ to steel pipe to and from ______ ______.

15. Cleanouts, Test Tees

a. Cleanouts at end of each horizontal drainage run or change in direction at each branch connection, at intervals not over 40 feet in horizontal runs. Each cleanout provided with brass ferrule and brass screw plug, chromium plated cast brass handhole ______ ____________ in finished room. For under flue or drains, __________ branch brought to finished floor level with long sweep 1/4 or two 1/8 bends, brass screw plug with counter sunk spud set flush and level, ______, in larger lines, ______.

b. All cleanouts shall be located and arranged and be easily accessible for rodding. Tests tees with brass screw plug ______ 30" above floor at foot of all drainage.

16. Hanger, Supports

a. Support piping firmly, prevent sagging, ________, ________ or ________, making due allowance for structural requirements. Suspended pipe shall be held by ________ ________ ______ ________ ______ ________ . Metal pipe covering protectors where necessary to prevent deformation of insulation. Chain, wiring or perforated hangers will not be acceptable. ________ ________ may be used in lieu of separate hangers, spacing per smallest pipe. ____________ pipe shall be supported by chromium plated cast ________ supports. Hook-plates for pipe supported from side walls. Support vertical waste vent, or drainage pipes at ________ ______ ________ . Install hangers in line adjacent to __________ and in accordance with the following schedule, except hub and __________ pipe 5' or less in length supported 5' centers, support close to hub. Vertical piping supported each floor.

Pipe Size, inches *Minimum Spacing Hangers, feet*

17. Floor, Wall, Ceiling Plates

a. Uncovered piping passing through floors, finished walls or finished ceilings fitted with __________ or ________ large enough to completely close openings around pipe, securely held in place, ________ __________ . Caulk water tight around pipe in unfinished room.

18. Sleeves, Inserts

a. Provide __________ secured in place, sufficient size to accommodate pipe passing through ____________ or ________ . Sleeves in structural supports, galvanized steel pipe. Elsewhere, either galvanized sheet metal or especially designed fiberboard sides with metal caps and washers, ______________________ . Sleeves through floors flush with finished floor and space between pipe and sleeve caulked water tight. Contractor to ___ ________________________________ .

19. Supports, Fastenings

a. Fixtures, ____________________, to masonry ____________________, ____________________, or ____________________. Exposed portions of fastenings shall be chromium plated. Support fixtures hung from __________ walls with concealed ____________________ attached to iron fish-plated at rear of walls, ____________________, __________ and ____________________.

20. Traps

a. Each fixture, all equipment requiring drain connection, equipped with trap same __________ ____________________, placed as near __________ as possible. Water closed traps integral with ware.

21. Sterilizing

a. Cold, hot water lines, storage tank and equipment, sterilized with sufficient chlorine to provide dosage not less then ____________________, contact period __________, all valves opened and closed ____________________ during __________. Following ____________________, water thoroughly flushed from system until residual chlorine content is not more than ____________________.

22. Cleanup, Adjusting

a. All parts of work left clean, ____________________, __________, ____________________ and __________ cleaned of grease and metal cuttings. Any discoloration or other damage to portions of building, its finish or furnishings, due to Contractor's expense. All automatic control devices adjusted for proper operation. All surplus materials and rubbish shall be removed as it accumulates. All equipment shall be left in safe and proper operating condition.

23. Tests

a. Drainage and venting system shall be ____________________ and ____________________ ____________________. If necessary to cover pipe before roughing in ready for inspection, tests shall be made with water at ____________________. Cold, hot water pipe tested ____________________, proved tight.

24. Insulation

a. All work shall be performed by an experienced insulation contractor. All piping shall be __________, __________ and ____________________. Sectional covering secured in place with outward ____________________ of sufficient number to properly secure the covering, ____________________ minimum. All open ends of pipe insulation shall be neatly finished off with ____________________. Contractor shall be required to remove any insulation showing evidence of deterioration and replace with new insulation as directed. Exposed piping in rooms, including equipment rooms, shall have ____________________.

b. Domestic Cold Water:

c. Domestic Hot Water Lines:

25. Guarantee

a. This Contractor shall guarantee all materials, workmanship, and the successful operation of all apparatus installed, and to furnish service free of charge on all portions of the systems, for a period of ______________________ of acceptance of construction, and shall repair any defect, or replace apparatus within reasonable time after notice thereof at his own expense. Provided such defect is in the opinion of the ______________ or ______________ a fault of the equipment and not due to the misuse of the equipment.

26. Pipe and Fittings

a. Drainage pipe installed __________ or more outside building lines, ______________________.
Drainage and vent pipe from ____,
______________. Asphaltum coated conforming to ______________________ except fixtures wastes above ground and ______________ and less in diameter and branch vents unless otherwise shown, may be galvanized steel pipe with ______________________________.
b. Water pipe above ground, hard drawn copper tubing, ______________, with __________, ______________.
Water pipe not laid in trench with sewer or drain pipe.

27. Water Valve Stops

a. ____
____
______________. All valves ______________ and less shall be ______________________: and smaller may be ______________, all others ______________. Check valves of ______________ swing type or angle plated loose key stops on supplies to all fixtures.

28. Fixture Connections

a. Floor supported water closets secured to heavy cast brass floor flange with ______________________,
floor flange anchored to floor with ______________________________. Joint between bowl and piping shall be sealed ______________ and ______________ tight by graphite asbestos gasket.
b. Exposed traps and supply pipes for all fixtures and equipment shall be connected to ______________________ unless otherwise specified.
c. Water connections to individual fixtures not less than the following:

29. Fixtures and Equipment

a. All fixtures designed to prevent backflow of polluted water or waste into water supply system. Plating may be of ________ ______________ or ______________ on polished surface. All fittings on fixtures shall be same surface finish. Flush valve handles shall be ________________. ________________ stops and supplies with plated escutcheons for all fixtures. Mounting heights of lavatories, drinking fountains and sinks shall be as directed. Immediately after fixtures are set, cover with ________________ paper glued on; also assure trim is adequately protected. Provide guards and boxing as necessary to protect fixtures from normal operations of other trades. Before installing fixtures, all water pipe shall be thoroughly flushed out to remove all dirt, oil, chips or other foreign matter. Upon completion of the work, this Contractor shall remove protective covering from fixtures and thoroughly clean and polish fixtures and trim. All fixtures shall be ________________ or equal.

1) Lavatory

2) Water Closet

3) Urinal

4) Sink

5) Service Sink

6) Shower

30. Floor Drains (FD)

In floors on earth, with trap, integral or adjacent cleanout, leave cast iron body with brass grate, threaded brass cleanout plug, spigot outlet. Floor drains in finished rooms to have ______________________________ and in unfinished rooms to have ________________. Indirect waste drains. ______________________. All brass floor drains with funnel strainer, provide drains with special offset rims ____________________ where required for floor coverings.

31. Wall Hydrant

____________________ or equal, cast brass ______________ non-freeze wall hydrant, ________ hose connection, polished brass face, key handle and brass wall sleeve fitted with brass locknut.

32. Drinking Fountain (DF)

________________ recessed unit with non-squirting bubbler head, self closing control valve adjustable for continuous flow. Automatic volume regulator screw driver regulating stop. ________________ connection strainer, brass trap with cleanout extension to wall.

33. Roof Drains

Roof drains, ________________ as indicated on plans and as recommended by manufacturer.

34. Water Heaters

a. Furnish and install ________________ heaters as shown. Furnish with ________________ thermostat and connect heating elements through contactor. Contactor shall be ________________, ________, ________________. Each unit to be ________ at ________ three phase. Units are ________ high ________ diameter with ________ outlet and inlet connections.

b. Hot water storage tank, ________ diameter ________ long ________ gallon capacity, working pressure of ________________. ________ inspected and labeled ________ approved. Tank heads ________ flange quality steel tank sheet ________ flange quality steel. Galvanized inside and out after fabrication and test. Tank to be insulated after installation with ________ joints and staggered, secured with galvanized annealed steel wire over insulation ________ hexagonal mesh wire stretched and ends tied together. Finish with two coats ________________. First coat to dry before second coat is applied. Cover with ________ canvas cemented with ________________ ________, ________________. Paint with ________ paint. Tanks to have hubs as necessary to complete the plumbing installation. Contractor to install with ________ ________ temperature, and pressure relief valve, shutoff valve at tank inlet, and ________ Drain valve.

35. Kitchen Equipment

a. All kitchen equipment will be furnished by others, including traps, valves, strainers, or other components necessary for installation. This Contractor shall rough in for all equipment and make final connections to the following items:

Information for all equipment (kitchen) will be furnished by the General Contractor.

36. Circulating Pumps

a. Provide ________ ________________ pumps all with flanged in line fittings, two will be ________ flange and one ________ flange each will be ________ single phase motors. Two pumps will be mounted on the supply of the ________________ ________, and shall be adjusted for ________ degree. One pump shall be mounted on the return of the ________ ________________ ________ and controlled by a ________ mounted on the supply line, which will be set at ________ degrees F. Contractor shall furnish aquastats.

Name ______________________________ Date ______________________________

DIVISION 35A–HYDRONIC HEATING

WORKSHEET 44

1. Scope

The work under this heading shall include furnishing and installing all materials and equipment specified and shown on the drawings for the Heating System.

2. Gas Piping

a. *Material*

All gas piping shall be __________ __________, __________ __________, __________ __________. Pipe __________ and __________ shall be __________. All screwed fittings, except cocks and valves shall be __________ weight, __________, __________ __________ __________. Valves shall be __________, __________ __________, __________. All underground piping shall be __________ __________ __________.

b. *Installation*

All piping shall be __________ after cutting and all screwed joints shall be made up with __________ applied to the male threads only. All pipe shall be run true to line without pockets and with even pitch to a suitable point where an approved drain cock shall be provided.

No unions shall be used in concealed piping. All outlets not connected to equipment or appliances shall be closed with __________ __________.

Upon completion, the entire gas piping installation shall be __________ under and proven __________ __________ an air pressure of __________ pounds gauge per square __________.

3. Steam and Condensate Piping

a. *Material*

b. *Installation*

1) *Steam Mains* shall be run as shown on the drawings and shall be evenly pitched not less than ________ in ________ feet in the direction of ________ ________. The mains shall be of ________ indicated and shall ________ as shown on the drawings. Use eccentric reducers where reduction of main sizes occurs in the direction of steam flow in horizontal mains. The end of steam main shall be ________ into return line through ______.

2) *All Return Mains* shall be of ______ shown and shall be ____________________ not less than ________ in ________ feet in direction of flow.

3) *Allowance for Expansion* shall be made in the ______________ of all piping so that the usual variation in temperature will not cause undue stress at any point. Pipes shall be securely anchored where necessary to properly distribute expansion stresses.

   All branch mains shall be supported in such a way as to permit ____________ and ______________ and to relieve runouts of all ____________.

4) *Joints*—All pipe ________ or ________ shall be welded as set forth in the ______________________________.

   Piping ______ and smaller shall be __________.

5) Where pipe is welded, flanges shall be installed at __________ __________ and as required to dissassemble piping in equipment rooms for maintenance work.

6) Valves, traps and specialties for the steam piping shall be as specified hereinafter.

4. *Valves*

The Contractor shall furnish and install all valves of the sizes, pattern and make indicated on the drawings and described in these spceifications. All valves ________ and smaller shall have ______________. All stems shall be properly __________. Valves ______ and smaller shall be ____________. Valves larger than ______ shall be __________, iron body type, unless otherwise specified. Valves ______ and larger shall be ______________.

a. *All gate valves 2" and smaller* shall be

b. *All gate valves 2½" and larger* shall be

c. *All globe valves* shall be

d. *All angle valves* shall be

e. *All check valves* shall be

5. ***Traps***

a. *Thermostatic Traps*

Thermostatic traps shall be provided for all finned tube radiation, similar and equal to ______________________ ______________________, sized for the capacities of each unit as indicated on the drawings.

b. *Low Pressure Float and Thermostatic Traps*

Low Pressure Float and Thermostatic Traps shall be provided at __________ or ______________________ and at all ______________________, similar and equal to __________________. Traps shall have minimum capacities shown on the drawings at ______________________ pressure difference.

6. ***Strainers and Unions***

a. Strainers shall be

b. Unions shall be

7. ***Pressure Gauges***

a. Provide pressure gauges, U.S. Gauge, or approved equal, with white dial and black scale, size _________ dial. These gauges shall be located for easy reading. Each gauge shall be equipped with an __________ or ________ ______________ and shall be connected by means of a brass pipe and ____________ containing a ______________ ___________.

b. Steam system pressure gauges shall be as follows:

8. ***Boilers***

Furnish and install steel, fire-tube steam heating boilers as shown on the drawings. Boilers shall be arranged to ________ __________ hereinafter specified. Boiler shall be ___________ designed and stamped for a ___________ steam working pressure. Boiler shall be ________________ with ________________ doors and ______________________________.

a. Boiler shall be installed complete with the following items:

1) Burner shall be designed to burn ____________ efficiently, shall be completely automatic and shall be a complete ______ with ____________ and ____________. Burner shall be installed in strict accordance with ____________ recommendations and the requirements of the ____________.

2) Burner controls shall consist of an ____________ to cycle the burner operation, an air switch wired through the burner control circuit to hold the gas valve closed, unless the switch has been actuated by the burner fan. A high limit control to stop the burner and close the gas valve if the steam pressure reaches a predetermined high limit. Ignition shall be ____________, intermittent. Flame failure protection shall be ____________, ____________ or equal with ____________ and ____________ supervision of main and pilot flames.

   All control wiring, line and low voltage required for burner operation shall be furnished and installed by this Contractor.

9. Boiler Foundations

Each boiler shall be set on a ____________, isolated from the floor by an asphaltic expansion joint. Top edges of pad shall be chambered. ____________. The foundation shall be installed as recommended by the ____________ manufacturer. Pad shall have No. ______ reinforcing bars on ______ centers both ways. Pad shall be ____________ thick.

10. Fin Tube Radiation

Furnish and install fin tube radiation of the type, lengths, dimensions and capacity shown on the drawings and herein specified. All components of the fin tube installation, except the heating elements shall be given a phosphatized or bonderized treatment to prevent ______ and shall be finished with a ______, ______ ______ ____________. Radiation shall be installed in strict accordance with the manufacturer's recommendations. The fin tube radiation shall be manufacturer indicated or equal. Installation shall be complete with ______, ______, ______, ______, ______, ______, ______ and ____________ as required for a neat installation.

a. Fins shall be

b. Hangers shall be

c. Covers shall

d. All length of fin tube element assemblies exceeding ______ feet in length shall have

e. Fin tube ratings shall be approved under the

f. Delete fin tube radiation in classrooms for alternate __________.

11. Convectors

Convectors shall be of ________________ shown or equal. Convector cabinets shall be of ____________________ ______________________________________, and of type shown on drawings.

a. Cabinets shall have ________________ and shall be ________________.
b. Convectors shall be rated in accordance with ____ ____.
c. Convector elements shall be ______________, ______________________, ______________________, ____.
d. Convectors shall be installed as shown on the drawings with ______________________.

12. Propeller Unit Heaters

Furnish and install propeller type unit heaters of manufacturer shown or equal as indicated on drawings.

a. Heaters shall be enclosed in ____ ____.

Heating coil shall be ____ ____.

Fans shall be ____ ____.

b. Heaters shall be rated in accordance with ____.
c. Capacities shall be as indicated on drawings.
d. Horizontal discharge heaters shall have ________________ for both ______________ and ____________________ directional air control.
e. Control shall be as indicated in "Temperature Control."

13. Cabinet Unit Heaters

Furnish and install, where shown, cabinet unit heaters of manufacture shown or equal.

a. Units shall be ______________ mounted as shown with ______________, ______________ and ________________ as shown in the schedule. Units shall operate ____________.
b. Install valves and accessories as detailed.
c. Heaters shall be enclosed in ____ ____.

Heating coil shall be ____ ____.

Motor shall be ____ ____.

d. Control shall be as indicated in "Temperature Control."

14. Expansion Joints

Expansion joints shall be ______________________________ designed for use in ______________ and shall be suitable for installation in ______________ or ______________________ as required. Joints shall be installed where shown on the drawings.

15. Gas Vents

a. Boiler and water heater vents and breachings shall be ______________________________ ______________________________.

b. Installation shall conform to all recommendations of the ______________ and the ______________ ______________. Vents shall be ______________________________ and ______________ at the roof line and shall have approved type weatherproof caps.

16. Heating and Ventilating Unit

The unit shall be furnished and installed as indicated on the drawings.

a. Unit shall be complete with __________, __________, __________, __________, __________, __________, __________, __________.

b. *Casings.* Each air handling unit shall be housed in a casing constructed of ______________, not less than ______________________. The cabinet shall be adequately reinforced and stiffened with steel angles or other structural members, and shall be provided with all necessary interior panels, ______________ and ______________. Casing openings connected to ducts shall be equipped with ______________ and ______________ for attaching canvas or other flexible connections. All interior surfaces of the casing shall be rendered ______________. The cabinet shall be insulated on the inside with not less than __________ of ______________ ______________, ______________, ______________ insulation. Removable panels in the casing shall provide easy access to all parts for lubrication and servicing when the unit is installed.

c. *Heating Coil* unless otherwise specified, shall be ______________________________ ______________________________. The coils shall have been tested hydrostatically and proved tight under a gauge pressure of ______________. The coils shall be properly pitched to permit complete drainage and shall be encased in a ______________________. The coils shall have capacity not less than that indicated on the drawings. Coil shall be ______________________________.

d. *Circulating Fans* in unit assemblies shall be of the ______________, ______________type. Each fan unit shall have an air capacity not less than that indicated on the drawings. Each fan shall be installed complete with ______________________________. The fans shall be rated and constructed in accordance with the ______________. Fans shall be ______________ and ______________ balanced at all speeds. The fan shall have __________externally mounted self aligning, grease lubricated, __________. The fan shafts shall be made of ______________, and shall be provided with __________ __________and keys for the impeller hubs and fan pulleys, or with other equally positive fastening. V-belt drives shall be designed for at least ______ percent overload capacity. Each fan motor shall be equipped with ______________________________ or ______________________. Motor mountings shall be on a ______________________________ ______________________________.

e. *Motors* shall be ______________, ______________________________ as indicated in the schedule. Starter will be furnished and installed by the Electrical Contractor. Wiring to the motor lugs by the Electrical Contractor.

f. *Air Filters.* All filters shall be ____________ thick throwaway type ______________________________________.
Size and capacity to be coordinated with the units.

g. *Mixing Box.* Combination filter and mixing box shall be installed as indicated on the drawings. Each opening shall be provided with a ____________________________ – See "Temperature Control." Mixing box shall be factory assembled unit.

h. *Control of Unit* shall be as specified in "Temperature Control."

i. Unit shall be provided with the necessary piping, valves and accessories, as indicated on the drawings and as specified.

17. Unit Ventilators: (Alternate M-2 Only)

Furnish and install complete factory assembled unit ventilators. Units shall be of size, capacity and manufacturer shown on the drawings.

a. *Casings* shall be constructed of ____.
Casings shall be ____.

b. *Ratings and Capacities* shall be as indicated on the drawings. Units shall be arranged for the control cycle specified in Section "Temperature Control." Unit ventilators shall be tested and rated in accordance with the ASHRAE Standard Code for Testing and Rating Unit Ventilators.

c. *Fans* shall be _______________________, _______________, __________________. Fans shall be provided in sufficient number to insure __________________ operation with a maximum noise rating of ____________________.
Any unit found in the opinion of the Architect or Engineer to be excessively noisy shall be removed and replaced at the expense of this Contractor.

d. *Coils* shall be ____________, ____.
Capacity shall be in accordance with schedule shown on plans.

e. *Motors* shall be ________________, ____________________, __________________________________
____
____.

f. *Filters* shall be _____________ thick, renewable ____
____.

g. *Electric Wiring* for units shall be by Electrical Contractor to junction box inside unit. Interior wiring to be by this Contractor. Starting switch will be furnished by unit manufacturer.

h. *Protective Covering.* All unit ventilators, and accessory items shall be shipped with a ______________________ so applied that access panels can be removed without disturbing this cover. This Contractor shall be _______________ for _______________ to the satisfaction of the Architect or Engineer any damage incurred during installation.

i. *Fresh Air Intake Louvers in Masonry Walls* shall be ____
____. Furnish to General Contractor for installation.

j. Provide one unit ventilator under each classroom window.

18. Condensate Return Pump

Furnish and install one duplex condensate return pump with capacity and characteristics shown on the drawings. Pump shall be of manufacturer shown on drawings, or equal. Pump shall be driven by open _____________________________
____. Pump shall be capable of indicated operation while handling condensate to _______________. The equipment shall include _______________________________________
____
____.
A gate valve and check valve shall be installed in each pump, and a gate valve on the inlet to the receiver. Provide a pressure

gauge in the common discharge line. The receiver vent shall be ____

____. Vent may be connected to a common vent if so shown.

19. Water Feeder

Furnish and install on each boiler a combination water feeder and low water cutoff switch of a capacity equal to the evaporation rate of the boiler, ______________________________ or equal.

20. Boiler Water Treatment System

Furnish and install for the boiler feed system a pot type feeder having a minimum capacity of ____________ and constructed for an operating pressure of __________. Unit shall be complete with __________ and ____________ valves, ______________ valve, and ______________ valve. Feeder shall be installed on the ____________ side of the ____________________. A shop fabricated feeder construction from wrought iron may be used if approved. All piping shall be wrought iron with cast iron fittings.

21. Cleaning of Boilers and Piping

After the hydrostatic tests have been made and prior to the operating tests, the boiler shall be thoroughly and effectively cleaned of foreign materials. The boilers shall be filled with ______________________________________

____

____.

Name ____________________ *Date* ____________________

DIVISION 35B–VENTILATING WORKSHEET 45

1. Scope

The work under this heading shall include furnishing and installing all materials hereinafter specified or shown on the drawings, for the ventilating system.

2. Roof Exhausters–Centrifugal Type

Roof exhausters – ____________ shall be installed as shown on the drawings and shall be of the ____________, ____________________________ . Exhaust fans shall be of the manufacture shown on the drawings, or equal. Fans shall be of the ________, ____________________. ____________________ ____ ____________. Motors shall be ______________________________ .

Housing shall be fully ____________, and all __________ parts exposed to __________ shall be given a sprayed on insulating undercoating. Outlets shall be provided with removable ____________ . Units shall be mounted on approval vibration isolating bases.

Fan bearings shall be ______________ and provided with adequate and accessible means of lubrication. Each unit shall have a disconnect located in the motor compartment. Fans shall be ____________ in operation. Fan housing shall be easily removable for access to all ________________ .

3. Bathroom Exhaust Fans

Furnish and install residential type bathroom exhaust fans where indicated on drawings. Fans shall be of the ________ ______________ as shown on drawings. Fans shall be quiet in operation. Fans shall have ________________ dampers. Fan shall be of capacity shown on the drawings and of manufacture shown, or equal.

4. Roof Curbs

Roof curbs shall be of ________________________ and designed for mounting ______________ and ________________ . Curb shall be of the sound attenuating type with insulated lining. Curb shall be designed for flashing to the roof in an approved manner. Curb shall have adequate means for mounting dampers where dampers are specified or shown on the drawings.

Name ________________________________ Date ________________________________

DIVISION 35C–HEATING AND/OR VENTILATING **WORKSHEET 46**

1. General

The work required under the heating and ventilating contract shall include all ________________ to completely install with proper trades the system ready for use as outlined in these specifications and as indicated on the Plans. All work, materials, and manner of placing material to be in strict accordance with latest requirements of the manufacturer, ________________________, and with local and state laws and ordinances relating to this work. All material shall be __________, __________, of ______________ in every respect, and shall conform to standards of __________, and established standards for the particular type of equipment or material.

2. Drawings

Drawings forming a part of these specifications are numbered and described as follows:

Sheet No. __________

3. Data and Measurements

Data herein and on drawings is exact as could be obtained. Absolute accuracy not guaranteed. Obtain exact ________, ____________, ____________ etc., at site and satisfactorily adapt this work to actual building conditions; installing systems generally as shown or indicated.

4. Delivery and Storage of Materials

Provide for delivery and safe storage of materials. Deliver materials at such stages of work as will expedite work as a whole and mark and store in such a way as to be easily checked and inspected.

5. Materials and Equipment

a. When make and type definitely specified, __________ must be based on that make and type.
b. Approval granted on shop drawings and manufacturer's data sheets rendered as service only and not considered guarantee of quantities, measurements or building conditions; not construed as relieving Contractor of basic responsibilities under Contract.

6. Installation of Work

Examine drawings and specifications for other work, and if any discrepancies occur between plans for this work and plans of others, report such discrepancies to Architect or Engineer and obtain written instructions for __________ necessary to accommodate this work to work of others. Any changes in work, made necessary through neglect or failure of contractor to report such discrepancies, made by and at expense of ____________. Confer and cooperate with others on work and arrange this work in proper relation with theirs. Contractor held solely responsible for proper size and ____________ of __________, __________, __________, __________, __________, ______________ etc.,

required for proper installation of work. Arrange with proper contractor for building in anchors etc.; for leaving required chases, recesses, openings, etc; do all cutting and ______________ made necessary by failure or neglect to make such arrangements with others. Any cutting or patching subject to directions of Architect or Engineer and not started until approval obtained. Damage due to cutting repaired by this contractor. Notify general contractor exact ______________ and ________ ______________ all openings required for recessed units and equipment items.

7. *Electrical*

The Electrical Contractor shall provide all ______________ and ______________ to __________ all equipment and controls. The Heating and Ventilating Contractor shall furnish to the ________________________ all ______________ , ______________ or ________________ controls, ______________, ___________, ________________________ required for operation of the equipment specified, and furnish complete wiring diagrams and instructions within two weeks after contract award. Safety disconnect switches, where required, shall be provided by the Electrical Contractor.

8. *Duct Work*

a. Duct work must be permanent, rigid, non-buckling, and non-rattling with all flat sides cross broken. Joints in duct work shall be _______ ___________ . Galvanized iron sheets of the following ______________________ shall be used in the construction of all duct work.

   Greatest Dimension *U.S. Gauge*

   Diameter, inches *Gauge*

b. All changes in the direction of air flow and in cross sectional area shall be accomplished with well designed, properly installed __________________________________ . Supply ducts must be securely supported by ________________ , ________________ , _______ or ______________________ . No nails shall be driven through duct walls and no unnecessary holes shall be cut in them.
c. The size of the ducts shall be as indicated on the plans and shall be net inside dimensions of the sheet metal duct.
d. The general location of the ducts and flues shall be as indicated on the plans exact locations shall be determined at the building and the Contractor shall furnish and install such additional bends and offsets as may be required to bring the ____ __________________ into proper relation with other equipment and features of the building.
e. Where thermal duct insulation is called for on the Plans the insulating material shall be vapor barrier faced ____________ __________________________ of the thickness specified as manufactured by ______________________________ or approved equal. The insulation shall be held in place with spot __________ of a quicktacking, rubber-based adhesive on approximately __________ centers. All end and longitudinal joints shall be butted firmly and sealed by taping with a __________ wide vapor barrier pressure sensitive tape. Tape ends are to lap at least __________ . Plain vapor barrier tapes applied with adhesive may be substituted.

f. The ________________ round duct shall be a prefabricated glass fiber duct system of the size shown on the plans as manufactured by ____. The duct system shall be installed in accordance with the manufacturers' recommendations as outlined in their manual for installing ____________________.

9. Registers

All supply registers and return grilles shall be of the size and type shown on the Plans.

10. Gas Piping

11. L.P. Gas Storage Tanks

12. Guarantee

a. The Contractor, by accepting this specification and signing of the Contract, acknowledges familiarity with all requirements and guarantees that every part including equipment, fittings, ______________, ____________, ______________ etc., used to make up the system herein provided, will be of the highest quality in its field and will be erected by skilled and ______________ workmen in a thorough and substantial manner.

b. This Contractor shall guarantee all ______________, ____________, and the successful operation of all equipment and apparatus installed by him for a period of one year from the date of final acceptance of the whole work, and shall guarantee to repair or replace at his own expense any part of the apparatus which may show defect during the time, provided such defect is, in the opinion of the ______________ or ______________, due to imperfect material or workmanship and not carelessness of improper use.

13. Testing and Operating

a. After the plant has been completely installed the Contractor shall operate the entire system for ____________ days and will thoroughly clean and ________ the system to remove all dirt, sand, scale, grease, lint, etc. Before final acceptance. Before turning the work over to the Owner, the Contractor shall have all parts tested and in good working order and adjusted to operate as intended by the specifications.

b. All equipment bearings and motor couplings shall be lined up true and even so that there is no ___________, ________, or ____________present when the equipment is operating at full rated capacity. When aligned, all equipment shall be securely anchored in place.

c. The Contractor shall adjust all supply, __________, and _________________ to provide the amount of air called for on the Plans. Velocity readings shall be made with a __________, ________________, or other approved ____________ in an approved manner and the readings recorded as directed. Where the amounts of air are not shown on the drawings, the velocity through the openings shall be equalized. After preliminary adjustments are made, the system shall be checked with ________________________and supply ducts to each room adjusted so that all temperatures are equalized.

d. All thermostats shall be adjusted and set to operate as intended.

e. All air filters shall be clean.

f. The Contractor shall instruct the Owner on operation and maintenance of the system during the ________day operating period.

g. The Contractor shall furnish all ______________required for the __________; the Owner will furnish electrical current and fuel required for the __________. The Contractor shall thoroughly lubricate the system during the test and leave the system completely lubricated upon completion of the test and the system is accepted.

14. Plumbing

The Plumbing Contractor shall provide and install a floor drain for removing condensate from the _________________ located in the ____________________.

15. Equipment and Controls

a. The heating and ventilating units shall be of the model and size as shown on the Plans and hereinafter scheduled. Manufacturer's name and equipment numbers given in the following specifications are for the purpose of establishing __________ ____________, and equipment by another manufacturer, completely meeting such standards, will be acceptable.

b. Schedule of Furnaces:

1) Furnace:

2) Controls:

3) Fresh Air Intake:

4) Provide ______________________ as shown on the Plans.

a) Furnaces:

b) Controls:

c) Fresh Air Intake:

d) Provide ____________________________ as shown on the Plans

c. Schedule of Exhaust Fans:

Model	Make	Capacity CFM	Location	Remarks

16. *Combustion and Ventilation Air*

17. *Air Conditioning*

Name ____________________ Date ____________________

DIVISION 36–ELECTRICAL WORKSHEET 47

Generally the electrical contract is let out under a separate contract to an Electrical Contractor. Therefore, in this division, we again list the General Notes, etc. So that you can become familiar with all phases of the electrical contract, you are required to fill in the spaces left blank. List your materials and outline your installations in the required places. Be sure that you are listing only the materials and installations called for in your building.

Division 36–Electrical, in your Specification Textbook, will help you to complete this worksheet. Study it carefully.

Electrical

1. General Notes

a. The bidder shall carefully examine the __________ and ______________, and ________ the site of construction. The ________ shall fully inform himself to all the existing conditions and limitations, and shall include in his bid a sum to cover the cost of ______________________________________,

b. The ____________ shall be familiar with the city, county and state and national codes, and to which code takes precedance, and where the drawings and ____________________ do not cover such items the ________________ shall be ______________ for such.

c. It is pointed out that the general conditions and special conditions of the ________________________ are considered a part of these spceifications, and this __________.

d. Should bidder find discrepancies in, or omissions from the __________ and ______________, or should he be in doubt as to their meaning, he should at once notify the __________ _____ ______________, who will send written instructions to all bidders. Neither the Owner, Architect or Engineer will be responsible for any ________ instructions.

e. Provide all items, articles, equipment, material, operations, and tools for erections or mechanical systems shown on drawings or specified herein, including labor, supervision, and incidentals required and necessary to complete systems for successful operation. Install systems in complete accordance with standards of local or state codes governing such work. Place complete system in safe and satisfactory condition; adjust all automatic control devices in ____________________.

f. All equipment installed in strict accordance with ______________________ and these ______________ unless otherwise approved in writing. Vertical and horizontal dimensions of all fixtures, units, and equipment ________ for installation in spaces provided therefor. Ample clearance provided for maintenance and repair. Instruct __________ in operation and maintenance; ______________ one set to the ________ of repair parts list and one set instructions on ____________ and ______________ of all equipment installed. All material new and unused.

2. Fees

a. Pay all ________, __________, __________for inspection, ________ ____________ on connection, etc., in connection with the contract. Upon completion of ________ furnish owner with certificate of final inspection from ________ ____________having jurisdiction.

3. Drawings

a. Drawings forming a part of these specifications are numbered and described as follows:

4. Data and Measurements

a. Data herein and on drawings is ___________ as could be obtained. Absolute accuracy not guaranteed. Obtain exact ______ _________________, _______________, __________, _________, at site and satisfactorily adapt work to actual building conditions; installing system generally as shown or indicated.

5. Delivery and Storage of Materials

a. Provide for ___________ and _________ _____________ of materials and arrange with other contractors for introduction into building of equipment too large to pass through ____________ _____________. Deliver materials at such stages of work as will _____________ work as whole and mark and store in such a way as to be easily checked and inspected.

6. Materials and Equipment

a. When make and type definitely specified, ______ must be based on that make and type. When ______________ one of two or more equally acceptable make or types, one of make and type mentioned but that make or type throughout. When specified a certain make or type, "or approved alternate," make and type specifically mentioned unless written approval of "alternate material" is obtained from Architect or Engineer prior to delivery on job.

7. Approval Data

a. Approval granted on _______ ____________ and ___________________ _________ ________ rendered as a service only and not considered guarantee of quantities, measurements or building ____________; not construed as _________ contractor of basic responsibilities under contract.
b. For items _____________ from those specified herein and requiring approval, the _____________ shall submit complete data within 30 days after notice of contract award. Failure to submit ____________ _____________ data within the specified 30 day period shall be interpreted to mean that the Contractor will furnish equipment as specified. Requests for _____________ or ______________ ______________ after the 30 day period will not be considered unless it can be shown that, due to circumstances beyond the contractor's control, the specified material is not available and cannot be obtained in time to prevent unnecessary delay of construction.
c. For ____________ of approval the Contractor shall submit three sets of complete schedules of material and equipment proposed for installation; include catalogs, diagrams, drawings, data sheets and all descriptive literature and data required to enable the Architect or Engineer to determine compliance with construction and contract requirements. Schedules and all data _______________ at one time; no _______________ will be given to partial or incomplete schedules.
d. Include with data, _________ ____________ describing all differences between substitution and item specified; ______ ____________ changes required for piping, ducts, wiring, space, structures, and all portions of project ________________ by substitution.
e. The Contractor shall be limited to one approval on a singular piece of equipment. ________________ will not be given where the Contractor submits approval data on piece of equipment, by more than one manufacturer.

8. Installation of Work

a. Examine _____________ and __________________ for other work and if any discrepancies occur between plans for this work and plans for work of others, report such discrepancies to _____________ or ___________ and obtain written instructions for changes necessary to accommodate this work to work of others. Any changes in work, made necessary through neglect or failure of ______________ to report such discrepancies, made by and at _________ of Contractor. Confer and cooperate with others on work and arrange this work in proper relations with theirs. ______________ will be

held solely responsible for proper size and ________________ of anchors, inserts, ________, chases, recesses, openings, etc., required for proper installation of work. Arrange with proper contractor for __________ in anchors, etc., for leaving required chases, recesses, openings, etc., do all cutting and patching made necessary by ________ or ________ to make such arrangements with others. Any cutting or ___________ subject to directions of __________ ____ ______________ and not started until approval obtained. Damage due to cutting shall be repaired by the ______________. Notify General Contractor of exact sizes and locations of all __________ required for recessed units and equipment items.

9. *Tests*

a. Entire installation shall be free from __________ ________________ and __________ __________. Test to be made in the presence of the Architect or Engineer. Each panel to be ____________ with mains disconnected by the Architect or Engineer. Each panel to be with ______ ________________ from the feeders, branches connected and ______________ closed, lamps removed or omitted from sockets, and all wall switches closed. ____________ tested with the feeders disconnected from the branch circuit panels. Each individual power equipment is to be connected for proper operation. In no case shall the _____________ ______________ be less than that required by the ___________ ____________________.

10. *Rubbish*

a. Remove from building, premises and surrounding __________, ___________, ________, all rubbish and debris resulting from operations as it accumulates and leave all material and equipment and spaces occupied by them absolutely ___________ and ______________ for use.

11. *Construction Power*

a. Provide temporary electric service for use during _________________ complete with metering connections approved by the _______ _______ ______________. Provide by ____________ or _____________ connections, grounding type outlets, ______________ volts, _________ cycles, __________ _________, so located throughout the building that no more than ________ __________ of cable will be required to reach any portion of the building, and with the ground pole properly connected to ________ _________ ________ or __________ ______. Temporary power supply shall be inconnected to water service pipe or ground rod. Temporary power supply shall be installed in such manner as not to endanger _________ or ___________. Connections to outlets in the temporary power distribution system shall be made with a minimum of No. _______ conductor. Temporary lighting for construction purposes, and payment for all power will be by the General Contractor, temporary protection shall be ___________ amp at source of power.

12. *Grounding*

a. Identified neutral in the interior wiring system all ____________ ______________, ______________, _____________, and _______________ permanently grounded to the water system as near as practicable to the point of entrance with approved copper ground clamp. Area of contact at point where wires connect to be sufficient to provide a current carrying capacity equal to that of the wire. No wires smaller than No. _______ shall be used for ____________. Covering of identified neutral to be _______ ____________, or _______________ in another ________ ____________. Equipment grounds ___________ finished. Connections protected by ___________ ______________. Type and regulation of the ______________ _____ ___________ ________________ and the ______________ ____________________ ______________.

13. Wire and Wiring

a. Type-Wire for general light and power wiring, ________ or the ____________________, rate ________ volts, equal to that manufactured by the ________ ________ ____ ________ ________ No. ____ and larger, stranded Conductors soft drawn, annealed ________ with conductivity not less than ______. Insulation "Code" Type ________ for branch lighting and power circuits; type ______ for ________. Wire installed in conduit in floors on earth, and underground, ________ ________ type ______ or ______ unless otherwise noted. Neutral or grounded wire in two or multiple wire branch circuits white finished or otherwise distinctively colored. Multiple circuits coded in accordance with the ________ ________ ________. Deliver wire on job in original coils bearing manufacturer's name and ______ ____, and ________ ________.

b. Size. As required and in no case smaller indicated by diagrams. No wiring smaller than No. ______ accepted for line voltage circuits. Neutral same size as outside wires. On Homeruns where distance to first outlet exceeds fifty feet, minimum size No. ______. For low voltage control switching, and signal systems, the manufacturer of the system shall recommend the wire size to be installed.

c. Splices, taps and terminals. No splices or taps permitted except at outlet, ________, ______ and ________ ________. Wires No. ______ and ______ connected to ______ and ________ with ________ ________ or ________. Taps on or splices in wires No. ______ and ______ made with ________ ________. All lugs and connectors and sufficiently large to enclose all ______ of conductor and securely fastened. Splices or connections of No. ____ and ______ may be soldered or connected with ________ electrical ______ ________. Where spring connectors are used, insulate with ________ ________, soldered joints or be insulated with ______ and ______ or ______ ________. Insulation be equal to that of conductor. Prior to making connections, wires shall be ______.

d. Installation.

14. Raceways

a. Rigid conduit shall be of ____ ______, ________ ______, ________ ____ ________ ________ ________ ______ ________ and with ______ ______ ________ ________ ________ surfaces. Electrical metallic tubing of ______ ______, ____________, and with ____ ______ ______ ______ ______ ______ ______ ______. Elbows and fittings shall be of the ______ ______ as ________, delivered in ____ ______ ________, with maker's name and trademark and ________ ________ on each ______.

b.

c. Non-metallic conduits shall be ______ ____ ________ ______ ________ at each end, and the ____ ______ ________ ________ ______ a No. ______ ______ ______ ________ ________ ________ ____ ________. Adapter at the building end shall be installed ________ ______ from the point where the service enters the building, and conduit extended to service entrance equipment indicated on riser diagrams. Adapter at the supply end shall be installed within ____________________

____

____

____

____.

d. Service conductors from Pole to transformer slab furnished and installed by ________ ____________, to be code type ____________, sized as shown on plans.

e. Install as one complete system, with all joints in ________ ________ ______________ to boxes ____________ ________ ______________perfect. Cutting done with __________, and ________ ________ __________ ________ __________. Joints made with standard fittings and watertight. Joints in ________made with watertight compression type fittings. Indenter type fittings for __________not acceptable. Changes in direction made with trade elbow of the same size. All bends made smoothly with out crushing pipe or injuring protective coating. Not more than ________ __________ ______________, or equivalent, installed in a line between any two outlet boxes, cabinets or pull boxes. Pull boxes installed whenever necessary to comply with foregoing provision. Raceway to enter boxes squarely, and to be firmly attached thereto. Exposed raceway to be supported with ________ ________ or __________ __________ __________, equally spaced not more than __________ __________ apart. Raceway concealed in ______________ __________ ______________shall be erected ahead of construction, and wired in place. Exposed raceway run in a neat and slightly manner, square with walls and ceilings, adjacent runs parallel, at same level, and with concentric bends. Keep raceway systems free of ____________ ________ ____________, and swab clean before wires are pulled.

f.

15. Outlet Boxes

a. Provide at each current consuming or ____________ ________. Boxes set plumb and level, and secured firmly in place with face of box or ________ ____________ ______________ with finished wall or columns. Boxes secured to conduit with ____________ ____________ ________ ____________, and to tubing with ____________ ____________ ________ ____________ with insulating bushing in throat or fittings. Conduit or tubing offset as required to enter boxes used for pulling. Fixture outlet boxes shall have approved ____________ ________ ______________ __________ __________. Outlet boxes for systems other than power distribution system shall be as required by manufacturer of equipment installed.

16. Position of Outlets

a. Note: Prior to rough-in:
   1) Obtain approval from Architect for Engineer or outlet locations shown or changes required therein.
   2) Consult with mechanical contractor to determine where boxes are to set to avoid interference with mechanical equipment on walls or ceilings.

b. The right is reserved to change outlet locations shown, up to the time of rough in, without change in contract ________. Outlets centered with respect to __________, ____________, __________, __________. Where several outlets occur in one room, they are to be ________________ arranged. Outlets set plumb or horizontal, secured firmly in place. The face of the box extending to finished surface of wall, ceiling or floor as the case may be, without projecting or being recessed. Unless otherwise indicated or required, install outlets at heights noted below, dimensions being in ____________ above finished floor to center of device. Outlets shall be accessible for the intended use and shall not interfere with ______ ________, ____________, __________ ________ ______________ ______________, or any ____________ ________ ________________ as shown on plans for work of other contractors. All outlets shall be located to avoid installation of same at junction of dissimilar surface finished. Any outlet improperly located corrected at ________ expense.

*17. **Panel Boards***

a. Main type service distribution panel board of the dead front safety type equal to _____ _____ _____ _____. The _____ phase distribution panel board suitable for use on a _____ volt, _____ wire, _____ phase grounded neutral system. Each feeder unit properly identified. Panels arranged for size and number of circuits shown.

b. Other panel boards for use all _____ volt, _____ wire, _____ phase grounded neutral system panel boards similar to _____ with thermal-magnetic, quicktrip branch breakers. Branch breakers of the number and size shown. Branch breakers properly indexed, even numbers on the right, odd numbers on the left, and all active circuits indexed in an approved _____ attached to the inside of _____. Doors with combination lock and catch, two milled keys for each lock and all locks keyed alike.

*18. **Service***

a. Service, _____ wire, _____ phase, _____ volt, _____ cycles for _____ _____ _____ furnished and connected by _____ _____ _____ _____ _____ _____ _____ _____ _____ _____. Terminate service conductors at main service panel. Furnish and install metering connections and current transformer in Panel "p", ready for installation of _____ _____ _____ _____ _____ _____. Obtain detailed requirements of service arrangements from local power company and install in accordance with same.

*19. **Safety Switches***

a. Shall be _____ _____ _____ _____ at circuit voltage with number of poles as required by the circuit. Snap switches may be used for single phase motors of _____ or _____ and in accordance with _____. Enclosures to be in accordance with _____ for usage and location. Safety switches used unless noted. Each to bear manufacturer's name and rating and _____ _____.

*20. **Spare Circuits***

a. For recessed panel boards, _____ for future extension of the spare circuits shall be provided. _____ _____ shall extend upward or downward as directed and shall be _____ over the ceiling or under the _____ in such manner as to be easily accessible for extension to the future loads. One _____ conduit for each pair of _____ or _____ ampere single pole spares shall be provided.

*21. **Fixtures***

a. Furnish and install lighting fixtures, lighting equipment and lamps for all lighting outlets on drawings and listed in following fixture schedule, including connection of fixtures and equipment to electric wiring of building. Finish of fixtures to be manufacturer's standard finish except as otherwise noted. All joints in fixture wiring soldered and well insulated with _____ _____ _____ _____ _____ _____ _____, _____ _____ _____ _____ _____ _____ _____. All recessed fixtures to be prewired.

Note: List only those fixtures that are specified in your project.

22. ***Switches***

23. ***Receptacle Outlets***

24. ***Telephone System***

a. Furnish and install telephone entrances, terminal cabinet and underground conduit from pole to cabinet, and conduit from entrance to outlets shown. Provide outlet with stainless steel plates, bushing style. Telephone wiring will be installed by the ____________________, Contractor shall leave No. ________ iron pull wire in each conduit run. Confer with local telephone company as to exact requirements and location of service.

25. Fire Alarm System

If you are having a Fire Alarm System in your building, read your specification textbook and then fill in the following space provided.

26. *Program System*

If you are having a program system in your building, read this section in your specification textbook and fill in the following spaces provided.

27. *Sound System*

If you are having a sound system in your building, read the section on sound systems in your specification textbook and fill in the following accordingly.

Name ____ Date _______________________

DIVISION 36A–ELECTRIC HEAT

WORKSHEET 48

1. General

a. Electric heating equipment is based on ________________, as manufactured by ________________________ ____. Operational voltage shall be as per circuits on panel schedules. Wattage, lengths and types shall be as shown on plans, furnish all necessary items for complete operation and good appearance. All equipment shall be installed plumb and level. Furnish General Contractor all necessary information for installating lintels and providing proper insert openings for recessed equipment. Furnish all necessary contactors or control items necessary and connect equipment to circuit shown.

1) Base Radiation:

2) Counter Flow Units:

3) Forced Air Units:

4) Wall Radiant Units:

5) Room unit ventilators: Model ________ , Furnish and install where shown on plans. Air Capacities shown shall be based on the ASHRAE standard Air method of measuring. Unit Ventilators shall be installed in accordance with ____________________ ____________________ , and shop drawings. Units shall be designed and constructed to introduce a predetermined minimum quantity of outdoor air to the room during all periods of occupancy with up to ___________ room outdoors when rewelded, heavy gauge, __________________ steel with enclosures at each end for control and terminal blocks. Panels die-formed, ________ gauge ____________ steel, with ____________ ____________ rounded to __________ radius. Top grille and front panels removable for cleaning interior, too space ______ high. Cabinet to be _____________ deep and __________ high. Provide gaskets around outdoor air openings for airtight ________ _______ ___________ . Filter and Control access provided without the removal of front or top panels. Filter access strip shall cause a __________ ____________ switch to de-energize heaters, fan motors and close damper to outdoor air. Hard-baked, mar-resistant, acrylic enamel finish available in a manufacturer's standard color selected by the Architect or Engineer.

Steel,

Fans

_______ volt, single-phase shaded pole type with built-in automatic reset thermal overload protection, designed for continuous fan duty. Motors shall be ________________________________ connected to ______ _____________ and ______________ mounted to insure __________ _____________ ________ ___________, ____________ ______ _____ ___________ . Motors to be insulated from _________ _____ __________ _________ ________ _________ ______ ______ ______ ___________ . Motors shall be ______________________________ ____ ____

Motors shall be of ____ ____ .

Electric Heating Bank shall include ______________________ Fintube elements each with automatic reset snap-action thermal overheat limit switch set at 200 degrees F., all enclosed with heavy steel frame with end covers to protect _____________ . Each Fintube to be anchored at its center to assure noiseless expansion and contraction.

Helical wrap-on fins spaced _______ per inch furnace brazed to the steel sheath. Element surface temperature shall not exceed __________ under normal operating conditions. Built-in fan delay switch shall continue fan operation after elements are de-energized to dissipate any internal residual heat.

Each heating bank assembly shall be designed and wired for _________ volt, ________ phase power supply with single element switching.

Air Filter-one-piece filter for both outdoor and room air. Reusable permanent washable type, of __________ thick _____________________________ requiring no adhesive oiling. Filter frame to be a __________________ ____________________________________ .

Ventilation Control Damper Unit to have ___________ ___________ ___________ insulated to reduce heat loss when closed, arranged for airtight closure against silicone ____________________ _________ _______ ______________ , rotating on stainless steel shaft in Teflon bearings requiring no lubrication. Damper axis shall be ____________ mounted to _____________ __________ _____ _____ ________ _____ _______________ , preventing ________ through of out-door air without noise from additional flapper-type dampers.

Outdoor Air Intake ____________________ provide aluminum (Optional, clear anodized aluminum) intakes consisting of ________ gauge removable louver section with ________ square mesh aluminum bird screen. Intakes shall be furnished to ____ with instructions for setting, but electrical heating contractor shall be held ___________________ for _______________ installation.

Controls—See temperature control section.

6) Auditorium Unit Ventilator

If your project is an auditorium, read the section in your specification textbook on Auditorium Unit Ventilator and fill in the worksheet.

7) Furnish and install ____________________ Type ___________ Unit Heaters with heating and air delivery as shown on plans. Adjustable louvers and air-discharge fan guard and mounted at ________ points to absorb motor ______ __________________. Each unit shall have ________________ ________________ with night shutoff over ride switch.

8) Heating Cable

2. *Temperature Control*

These specifications cover the furnishing of a _____________ system of ___________ ______________________ controls as hereinafter described and as indicated on your drawings. Items not specifically shown or described but which are necessary for the successful operation of a complete system must be included.

The system shall be as manufactured by ______________________________ and shall be of the _________________ ________________________ type. The control system shall be under warranty for one year from date of acceptance by the Architect or Engineer.

All electrical wiring, including the low voltage wiring, shall be installed in conduit in accordance with the National Electric Code and local codes.

Wall thermostats shall have ______________________ and ______________________________ on cover.

Direct or immersion remote bulb thermostats shall have _____________ filled thermal elements and provided with ____ ________ type adjustments.

All modulating thermostats shall incorporate a feedback circuit from controlled device to thermostat to insure positive positioning.

(Internal feedback not acceptable; circuit must be complete from device to thermostat).

All motor operators shall be of the heavy duty _____________________________ and of the spring return failsafe type. Each motor operation shall incorporate a feedback circuit to thermostat controlling same coupled directly to the output shaft of motor to insure positive positioning.

All unit ventilator control equipment shall be factory mounted and wired by ______________________
Final adjustment and warranty of the control system shall be by the control manufacturer.

All field wiring necessary for temperature control system shall be installed by electrical contractor.

a. Control Panel:

b. Unit Ventilators:

c. Day Cycle:

d. Night Cycle:

e. Multi-Purpose Room:

f. Zone Control:

g. Unit Heaters:

3. Ventilation

a. Exhuast Ductwork:

b. Exhaust Fans:

c. Exhaust Registers:

d. Curbs:

4. Equipment Wiring

a. Furnish and install all disconnect switches required by ________________ and all junction boxes as indicated, with wiring to circuits shown.

5. Television System

6. Time Switch

7. *Exhaust Fans*

a. Exhaust Ductwork:

b. Exhaust Registers:

c. Exhaust Fans:

d. Curbs:

8. *Equipment Wiring*

a. Furnish and install all disconnect switches, contactors, and junction boxes indicated, with wiring to circuits as shown. All temperature controls, relays and other special controls for plumbing, heating and kitchen equipment. Wiring diagrams showing all necessary connections to such equipment shall be furnished by others. Final connections shall be such as to provide the operation specified under each item of equipment. Provide flexible metal conduit, stub conduits or other devices for completion of the equipment. The following equipment shall be connected as shown in panel schedules.

9. ***Alternate No. 1***

If one of your alternates applies to this Division, fill in the worksheet.